TRACE METAL REMOVAL FROM AQUEOUS SOLUTION

Special Publication No 61

Trace Metal Removal from Aqueous Solution

The Proceedings of a Symposium organised by the Industrial Division of the Royal Society of Chemistry as a part of the Annual Chemical Congress, 1986

University of Warwick, 9th—10th April 1986

Edited by

R. Thompson
Borax Holdings Ltd, Chessington

The Royal Society of Chemistry
Burlington House, London W1V 0BN

British Library Cataloguing in Publication Data

Chemical Congress (*1986 : University of Warwick*)
Trace metal removal from aqueous solution : the proceedings of a symposium organised by the Industrial Division of the Royal Society of Chemistry as a part of the Annual Chemical Congress, 1986, University of Warwick, 9th–10th April 1986.—(Special publication/Royal Society of Chemistry, ISSN 0260-6291; 61)
1. Solution (Chemistry) 2. Trace elements
I. Title II. Thompson, R. (Raymond) III. Royal Society of Chemistry. *Industrial Division*
IV. Series
541.3′422 QD541

ISBN 0-85186-646-8

Printed in Great Britain by
Whitstable Litho Ltd., Whitstable, Kent

Preface

Increasing industrialisation brings with it the twin problems of more thorough removal of unwanted and possibly toxic metals from chemical and nuclear process effluent, and the need to obtain some metals from progressively leaner initial sources or dilute recycle liquor. Legislation and economics link the two, but whatever the motivation the chemical principles are the same.

Traditional methods such as precipitation by liming, cementation or electrodeposition become less effective as metal ion concentrations fall to the low parts per million range and large volumes of liquor need to be handled. Resort may be made to ion-exchange and solvent extraction procedures, but resin degradation and evaporation losses can add to the problems of an unattractively high proportion of aqueous phase.

Alternative methods based on biological systems, with the complexation abilities of large molecules, the use of novel membranes and selective precipitants which overcome previous solubility product barriers are constantly under development. A symposium reviewing progress in these fields was considered a fitting contribution by Industrial Division to the Society's 1986 Annual Congress held at the University of Warwick. The eleven papers presented, including two on analytical techniques now available for metal determination at trace levels, are reproduced here together with a subsequently contributed item on carbon adsorption techniques for gold recovery.

Raymond Thompson
Immediate Past President, Industrial Division

Preface

Contents

Contents

Recovery of Heavy Metals by Immobilized Algae

By D.W. Darnall,* B. Greene, M. Hosea, R.A. McPherson, M. Henzl, and M.D. Alexander

DEPARTMENT OF CHEMISTRY, BOX 3C, NEW MEXICO STATE UNIVERSITY, LAS CRUCES, NEW MEXICO 88003, USA

INTRODUCTION

The binding of metal ions to microorganisms and the application of this phenomenon to water treatment are rapidly growing areas of interest. A survey of the scientific literaature reveals two distinct approaches to the problem: (1) use of living organisms and (2) use of a nonviable biomass. Metal-ion binding to living cells can occur either through surface adsorption or intracellular accumulation. This assertion is supported by work which showed that uptake of metal ions by live cultures cannot be accurately described by a model which assumes adsorption only to the cell surface.[1] Metal-ion binding to non-viable cells, however, is presumed to occur exclusively through surface adsorption.

A number of workers have investigated the feasibility of using actively growing algae in ponds or lagoons for wastewater treatment[2-10]. The basic approach has been to flow polluted waters through a lagoon in which an algal bloom is present. The effluent waters from such a system are then found to have lowered heavy metal ion concentrations. There are significant practical limitations to methods which employ living algal systems. Perhaps the most significant limitation is that algal growth is inhibited when the concentrations of metal ions are too high or when significant amounts of metal ions are sorbed by the algae.

Methods for water-treatment that employ non-viable cells are not complicated by the problem of attempting to maintain growth under adverse circumstances. In fact, Horikoshi *et al.*[11] found that heat-killed cells display a binding capacity for U(VI) three times greater than that measured for living cells. Instead, the biomass is treated merely as another reagent, a surrogate ion-exchange resin. The binding, or biosorption, of metal ions by the biomass results from coordination of the ions to various functional groups in or on the cell. These chelating groups--contributed by carbohydrates, lipids and proteins-- include carboxyl, carbonyl, amide, hydroxyl, phenolic, imidazole, phosphate, amino, thiol, and thioether moieties.

Tsezos *et al.*[12-16] studied the binding of U(VI) and Th(IV) to non-living *Rhizopus arrhizus*, a common fungus. They proposed that both ions are bound initially to amino groups present in chitin. The resulting complex then hydrolyzes, and forms insoluble hydroxy species which precipitate in pores on the cell surface.

Ferguson and Bubela[17] examined the biosorption of Cu^{2+}, Pb^{2+} and Zn^{2+} to frozen or freeze-dried preparations of *Ulothrix*, *Chlamydomonas* and *Chlorella vulgaris*. The degree of binding they observed was greater at pH 7 than at pH 3. They found that NaCl and $Mg(NO_3)_2$ inhibited only the binding of zinc, suggesting that selective adsorption of Pb^{2+} or Cu^{2+} was possible.

Nakajima _et al._[18] studied the binding of various ions to freeze-dried _Chlorella regularis_. They observed relatively selective accumulation of ions from an equimolar solution (1.0 mM in each ion at pH 5.0) which decreased in the following order:

$$UO_2^{2+} > Cu^{2+} > Zn^{2+} > Ba^{2+} = Mn^{2+} > Cd^{2+} = Sr^{2+}$$

The removal of UO_2^{2+} was unaffected by the presence of other ions, but Cd^{2+} uptake was strongly inhibited by equal concentrations of UO_2^{2+} or Cu^{2+}. The same authors also described a procedure for immobilizing _C. regularis_ in polyacrylamide for use as a chromatographic matrix.

If there is a single major conclusion to be drawn from our work to date, it would be that the surface of the alga, _Chlorella vulgaris_, is literally a mosaic of metal-ion binding sites--sites which differ in affinity and specificity. Both anions and cations can be bound. There are sites with high affinity for "hard" metal ions such as Al^{3+} and Fe^{3+}, and there are sites with equally high affinities for such "soft" ions as Hg^{2+}, Ag^{+} and Au^{3+}.[19] Selectivity can sometimes be gained by judicious manipulation of solution parameters. For instance, chromate/dichromate, bound negligibly at pH values around neutrality, can be bound completely at pH 2.0. This binding site diversity gives the algae a broad applicability not found in conventional ion-exchange resins. A second major advantage we have discovered with the algae is that, in contrast to many conventional resins, the cells have relatively little affinity for Ca^{2+} and Mg^{2+}. Thus, in hard-water treatment applications, the algae will be less prone to saturation by these non-toxic ions.

RESULTS AND DISCUSSION

pH Dependence of Metal Ion Binding.

We have found that metal ions can be divided into three classes based upon the pH dependence of binding to the algae. The first class is comprised of metal ions which are tightly bound at pH ≥ 5 and which can be stripped (or are not bound) at pH ≤ 2 (19). Many ions fall into this class: Al^{+3}, Cu^{+2}, Pb^{+2}, Cr^{+3}, Cd^{+2}, Ni^{+2}, Co^{+2}, Zn^{+2}, Fe^{+3}, Be^{+2} and UO_2^{2+}. The second class is comprised of metallic anions which display the opposite behavior of class I metal ions, i.e., they are strongly bound at pH ≤ 2 and weakly bound or not bound at all at pH values near 5. Ions in class II include $PtCl_4^{-2}$, CrO_4^{-2}, and SeO_4^{-2}. The third class of metal ions includes those metal ions for which there is no discernible pH dependence for binding and includes Ag^+, Hg^{+2} and $AuCl_4^-$. These three ions are the most strongly bound of all metal ions tested. Figure 1 illustrates data for the three classes of metal ions.

The data in Figure 1 were collected by incubating Chlorella cells (5 mg/ml) in 0.1 mM solutions of the metal ions in 0.05M

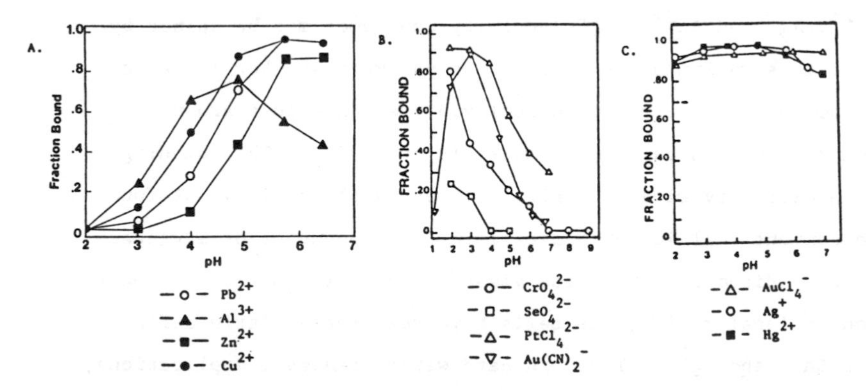

Figure 1. The pH Dependence of Metal Ion Sorption by Chlorella vulgaris

acetate buffer. The buffer was added to maintain accurate pHs between pH 4 and 6. Because acetate is a good ligand for many of the metal ions, even increased binding is observed in the absence of the buffer. Furthermore, complete binding of metal ions in all three classes is obtained when solutions are passed through columns containing the immobilized algae rather than by simply incubating the algae in metal-containing solutions.

We believe that the different pH binding profiles for the three different classes of metal ions are a result of the nature of the chemical interactions of each class of metal ions with the algal cells. The pH profile for Class I metal ions is consistent with the metal cation being the species which is bound to ligands on the cell wall. At pH values above the isoelectric point of the cells there is a net negative charge on the cells. The ionic state of such ligands as carboxyl, phosphate, imidazole and amino groups will be such as to promote reaction with metal ions. As the pH is lowered, however, the overall surface charge on the cells will become positive which will inhibit the approach of positively charged metal cations. It is likely that protons will then compete with metal ions for the ligands, thereby decreasing the interaction of metal ions with the algal cells.

The ions which form Class II are anionic in nature. We believe that the interaction of these ions with the algal cell is primarily electrostatic in nature. This is consistent with the fact that Class II ions do not bind at high pH where the overall algal surface charge is negative, but that they do bind at low pH where the algal surface charge will be positive.

Those metal ions in Class III are comprised of Ag^{+}, Hg^{+2} and $AuCl_4^-$. These metal ions, being "soft" in nature, prefer to form covalent complexes with "soft" ligands which contain the elements of nitrogen and sulfur. The reaction of these metal ions would therefore be expected to rather pH independent.

Table I summarizes data obtained from experiments such as those shown in Figure 1. The data shows the relative affinity of the algal biomass for different metal ions under a given set of conditions. At pH 5, Ag^{+}, Hg^{+2}, $AuCl_4^{-}$ and UO_2^{+2} are most strongly bound of all metal ions tested, while Zn^{+2} and Ni^{+2} are the more weakly bound in the acetate buffer system.

Table 1. Metal Ion Binding to Chlorella vulgaris

TEST METAL ION	% REMOVED	INITIAL CONCENTRATION
Au(III)	100	1.0×10^{-4} M, pH 2–7
Ag(I)	100	
Hg(II)	100	
U(VI)	100	1.0×10^{-4} M in 0.05 M Sodium Acetate at pH 5.0
Cu(II)	90	
Be(II)	80	
Al(III)	80	
Pb(II)	75	
Cd(II)	60	
Ni(II)	40	
Zn(II)	40	
Pt(II)	90	1.0×10^{-4} M, pH 2.0
Cr(VI)	84	

Figure 2 shows that some metal ions compete with one another for binding sites on the algae. Figure 2A shows the pH binding profile for nine different metal ions in the absence of one another, whereas Figure 2B shows the pH binding profile for the same metal ions when they are all simultaneously present at the same concentrations. While it is clear that some of the metal ions compete with one another for algal binding sites, it is of interest to note that the bindings of Ag^{+} and Al^{+3} are relatively unaffected by each other or other metal ions. This suggests that at least two classes of binding sites exist, one with specificity for hard ions such as Al^{+3} and others with specificity for soft ions such as Ag^{+}.

A

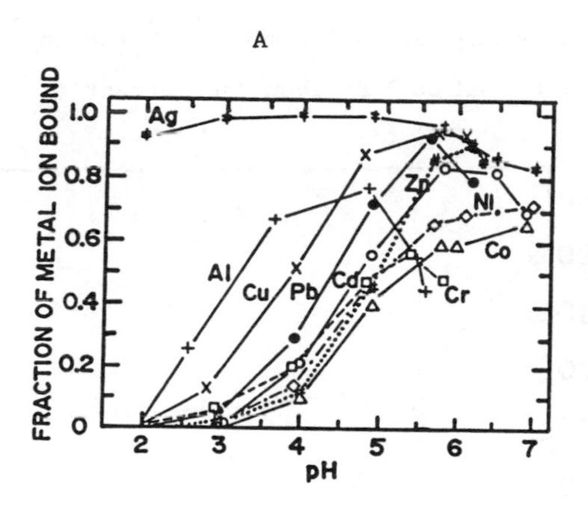

B

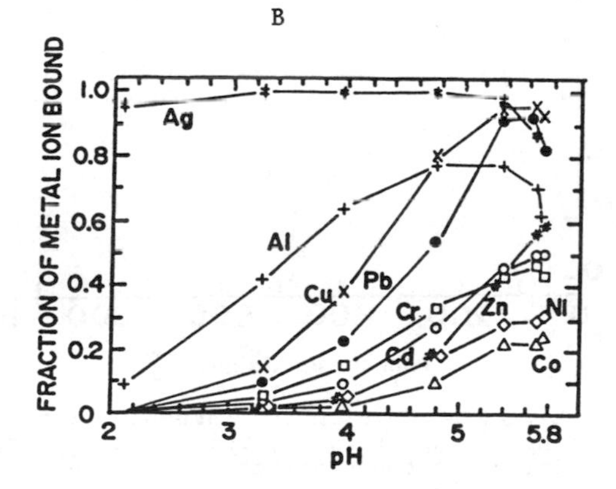

Figure 2. The Competition of Metal Ions for Binding Sites on <u>Chlorella vulgaris</u>. A. Algae (5 mg/ml) was reacted at the indicated pH values with separate solutions of each of the ions shown above at concentrations of 0.1 mM in 0.05 M acetate. After two hours the solutions were centrifuged and the supernatants were analyzed for the appropriate metal ion. The ordinate represents the fraction of metal ion removed from solution. B. This illustrates the same data as in A, except that all nine metal ions were simultaneously present during the experiment.

Initial experiments indicated that Ca^{+2} and Mg^{+2} were rather weakly bound by the algae. This suggests that the components of hard water may interfere minimally with the binding of transition metal ions to the algal system. That this is the case is shown in Figure 3 where it is seen that concentrations as high as 2000 ppm of Ca^{+2} and Mg^{+2} only inhibit the binding of 0.1 mM Cu^{+2} by about 30 percent.

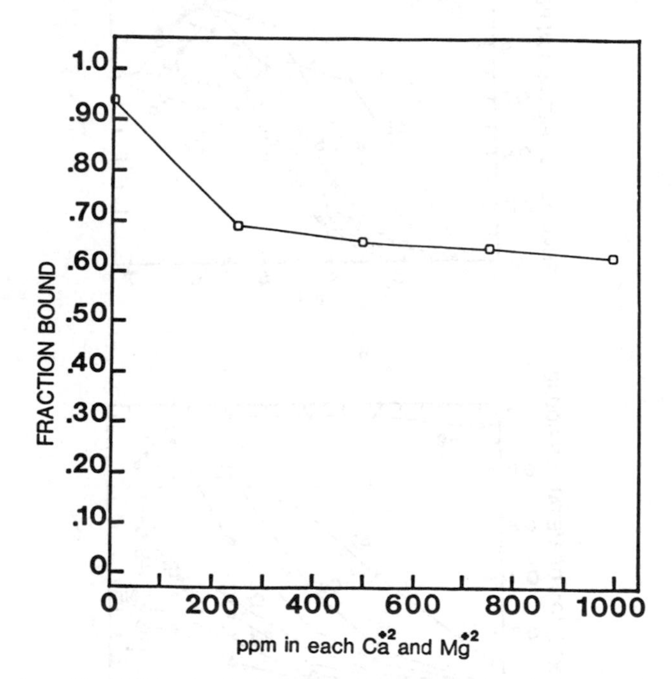

Figure 3. The Effect of Calcium and Magnesium Ions on the Binding of Copper Ion by <u>Chlorella vulgaris</u>. Conditions of the experiment were identical to those described for Figure 2A.

<u>Uranyl Ion Binding</u>.

The uranyl ion was found to be very strongly bound by <u>Chlorella</u>. However, when ground waters containing UO_2^{+2} were treated with algae, it was clear that an interference was present that inhibited the binding of UO_2^{+2} (Figure 4). Thus we

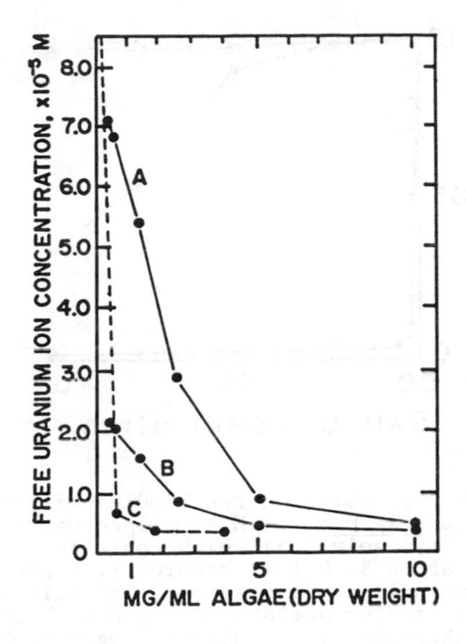

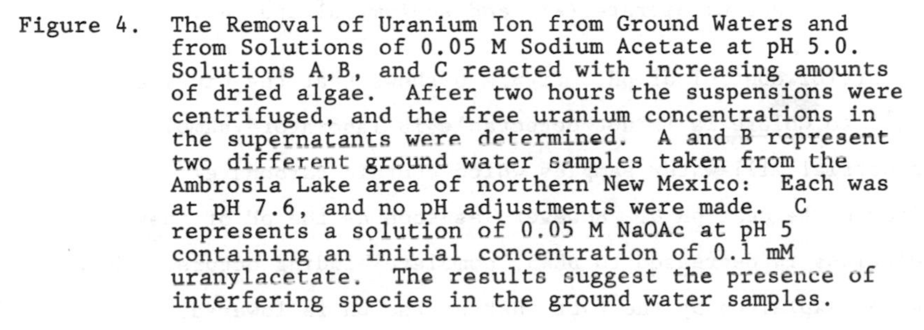
Figure 4. The Removal of Uranium Ion from Ground Waters and from Solutions of 0.05 M Sodium Acetate at pH 5.0. Solutions A,B, and C reacted with increasing amounts of dried algae. After two hours the suspensions were centrifuged, and the free uranium concentrations in the supernatants were determined. A and B represent two different ground water samples taken from the Ambrosia Lake area of northern New Mexico: Each was at pH 7.6, and no pH adjustments were made. C represents a solution of 0.05 M NaOAc at pH 5 containing an initial concentration of 0.1 mM uranylacetate. The results suggest the presence of interfering species in the ground water samples.

examined the effects of different salts on the binding of UO_2^{+2} to Chlorella. It was found that sodium acetate, sodium chloride, sodium nitrate and sodium sulfate were without effect on the binding of UO_2^{+2}, but that sodium phosphate inhibited the binding somewhat and sodium carbonate was inhibitory at even very low concentrations (Figure 5). Since the uranyl ion forms to strong complexes with carbonate, it was likely that carbonate was effectively competing as a ligand for UO_2^{+2}. The interference of carbonate can be completely eliminated, however,

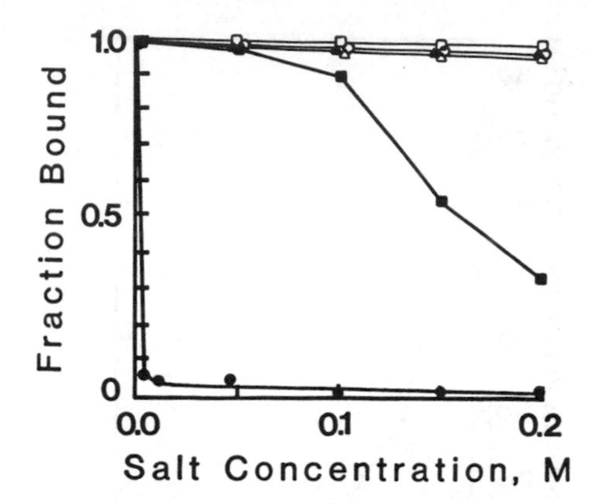

Figure 5. The Effect of Salts on the Binding of Uranium (VI) to Chlorella vulgaris at pH 8. Algae was suspended at 1.5 mg/ml in solutions which contained 0.05 M sodium acetate at pH 8, 0.1 mM uranium (VI) acetate, and a different salt concentration. The salt solutions used were sodium acetate (□), sodium bicarbonate (●), sodium chloride (○), sodium hydrogen phosphate (■), sodium nitrate (▲), and sodium sulfate (△).

by careful control of pH. Figure 6 shows that the binding of UO_2^{+2} to Chlorella is not detectable in laboratory samples or industrial mill-water samples which contain carbonate/ bicarbonate near pH 8. However, as soon as the pH of these solutions is decreased to pH 5, complete binding occurs. This can be explained by the fact that as the pH is lowered, carbonate and bicarbonate are protonated which leads to CO_2 being lost from the solution, and the carbonate interference is eliminated.[20]

The fact that carbonate was found to interfere strongly with UO_2^{+2} binding suggests that carbonate might be used to strip bound UO_2^{+2} from from algal cells. Figure 7 shows an experiment in which a 1.0 mM solution of UO_2^{+2} was passed

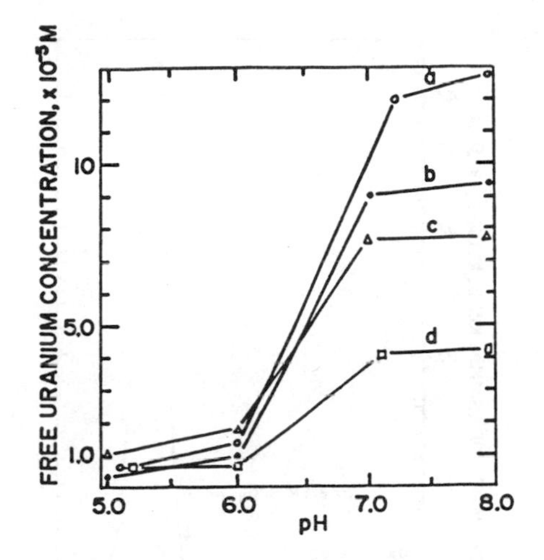

Figure 6. The Effect of pH on the Removal of Uranium Ion by Algae from Mill Waters and Sodium Bicarbonate Solutions. Algae suspensions in mine waters (curves a,c, and d) and 0.05 M sodium bicarbonate (curve b) were adjusted to the indicated pH values (original pH was 8). The suspensions were allowed to react for 2 hours, at which time they were centrifuged, and the amount of uranium ion remaining in the supernatant was determined. It is apparent that sodium bicarbonate strongly inhibits the binding of uranium ion to algae above pH 5 or 6. Algae-free control samples showed that pH adjustment alone did not result in any precipitation of uranium ion.

through a column containing Chlorella vulgaris immobilized in polyacrylamide. After passing 80 ml of the uranyl solution through the column and after extensive washing, no UO_2^{+2} was detected in the effluent. The bound UO_2^{+2} was then quantitatively recovered upon elution of the column with 0.05M $NaHCO_3$ at pH 8.0.

Gold Ion Binding.

Although the data in Figure 1 indicates that gold as $AuCl_4^-$ is bound independently of pH, the same is not true of all

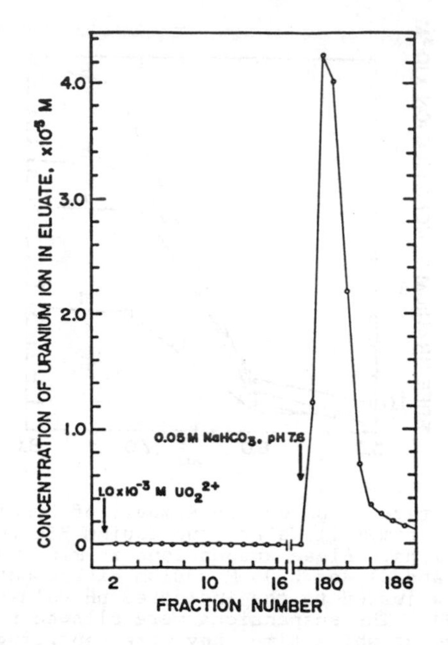

Figure 7. Binding of Uranium Ion to Immobilized Chlorella. A solution of 80 ml of a 1.0 mM solution of UO_2^{2+} in 0.05 M acetate, pH 6, was loaded onto a 20 ml column of polyacrylamide-embedded algae. The column was then washed extensively with 0.05 M NaOAc, pH 8.0. No uranium ion was detected in the eluate during either the loading or washing steps. Elution was then begun (fraction 178) with 0.05M $NaHCO_3$, pH 8.0. Fraction volumes were 5.0 ml.

complexes of gold.[21] Figure 8 shows the pH dependence of binding of various gold(I) and gold(III) complexes to Chlorella cells. The binding of gold(I) and gold(III) complexes of cyanide is strongly pH dependent with maximum sorption taking place at pH 2-3, The thiourea complex of gold(III) is bound weakly at pH 1-2 but is significantly bound at higher pHs. The thiomalate complex of gold(I), on the other hand, is strongly bound between pH 1-5, but is less strongly bound at higher pH values.

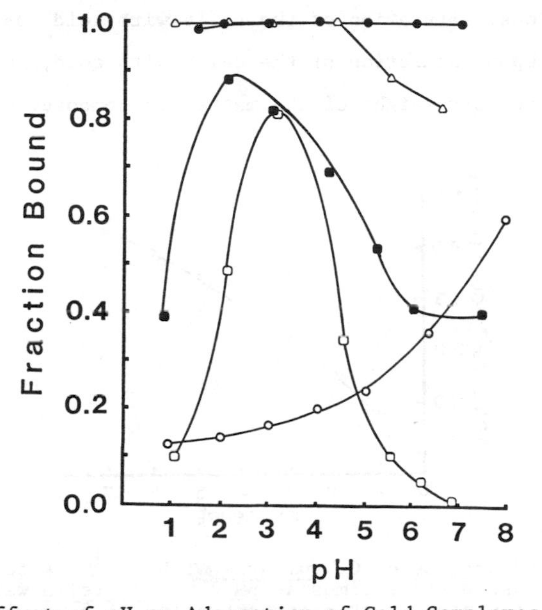

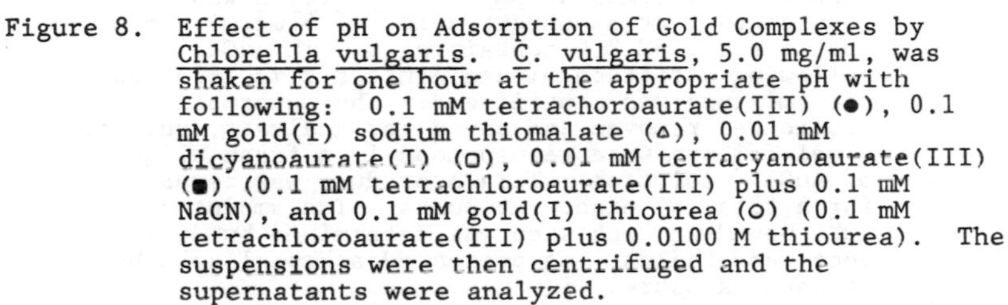
Figure 8. Effect of pH on Adsorption of Gold Complexes by Chlorella vulgaris. C. vulgaris, 5.0 mg/ml, was shaken for one hour at the appropriate pH with following: 0.1 mM tetrachoroaurate(III) (●), 0.1 mM gold(I) sodium thiomalate (△), 0.01 mM dicyanoaurate(I) (□), 0.01 mM tetracyanoaurate(III) (■) (0.1 mM tetrachloroaurate(III) plus 0.1 mM NaCN), and 0.1 mM gold(I) thiourea (○) (0.1 mM tetrachloroaurate(III) plus 0.0100 M thiourea). The suspensions were then centrifuged and the supernatants were analyzed.

The affinity of Chlorella vulgaris for $AuCl_4^-$ is stronger than for any other metal ion that we have tested. Our experiments indicate that $AuCl_4^-$ binding is independent of pH from pH 0-9.5. If C. vulgaris is added at a level of 5 mg/ml to a solution containing 0.1 mM $AuCl_4^-$, essentially complete binding of gold is observed. If these same algal cells are then exposed to a fresh solution of 0.1 mM gold, again complete removal of gold from the solution is observed. Figure 9 shows that even after eight such exposures of the algal cells to fresh gold solutions, saturation of the cells with gold has not occurred. Upon saturation of the cells with gold, more than ten percent of the dry weight of the matter is composed of gold.

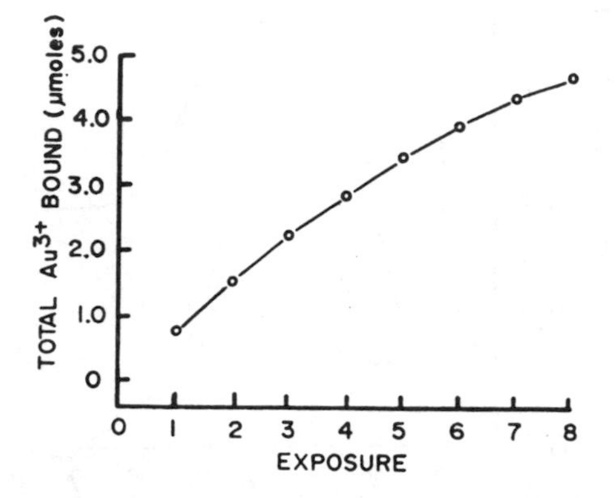

Figure 9. Saturation of C. vulgaris with Au^{3+}. After being washed three times at pH 2.0, Chlorella was resuspended, at a concentration of 5 mg/ml, in 50 mM acetic acid, pH 2.0 containing 0.2 mM $AuCl_4^-$. Following a fifteen minute contact time, the reaction mixture was centrifuged. The supernatant liquid was removed and saved for analysis, and the algal pellets were resuspended in a fresh aliquot of $AuCl_4^-$ solution. This procedure was repeated for a total of eight exposures. The amount of Au^{3+} bound at each step was determined by difference, and the data is presented as total gold bound at each exposure.

When C. vulgaris is incubated with $AuCl_4^-$, chloride analysis indicates that three moles of chloride ion are released for each mole of gold which is bound to the algae.[21] This data (not shown) is consistent with Au(III), which has a coordination number of four, being reduced to Au(I), which has a coordination number of two. This would mean that each bound Au(I) would be coordinated to one ligand derived from the algal cells and to one chloride ion. That this is indeed the case is verified by spectroscopic studies which show that algal-bound gold which is stripped from the cells is in the +1 oxidation state[21].

Interestingly, when C. vulgaris cells saturated with $AuCl_4^-$ are allowed to stand for 48 hours, the characteristic color of the algae changed from green to purple. This purple color was reminiscent of the "Purple of Cassius," the color exhibited by certain colloidal gold suspensions.[23] To confirm the presence of gold(0) on the algal cells, a visible absorption spectrum of colloidal gold was compared with the spectrum of gold-saturated algae (Figure 10). The spectra of colloidal gold and the aged algae-gold samples showed absorption bands near 525 nm, which were absent in the spectrum of freshly prepared algae-gold samples. The absorption maximum of colloidal gold varies somewhat from 525 nm, depending upon the size of the colloidal particles. This is because a large light scattering tail overlaps the absorption maximum. Nevertheless, these results are consistent with algal-bound gold being slowly reduced to gold(0). At the present time we do not know what is being oxidized concomitantly with the reduction of gold bound to the algae.

Plate 1 shows an electron micrograph of algal cells which were incubated with $AuCl_4^-$ for a period of several days. The plate clearly indicates that gold is clearly deposited along the cell wall as well as in the interior of the cell. The presence of tetrahedral gold crystals is also consistent with the presence of elemental gold deposited on the cells.[22].

We have accumulated evidence that there are at least three different classes of binding sites for gold (as $AuCl_4^-$) on Chlorella vulgaris.[22] The strongest of these sites is capable of binding gold(III) from very dilute solutions. For example we have been able to bind and recover 90 percent of the gold from a solution of artificial seawater that contained as little as 10^{-9} M $AuCl_4^-$.

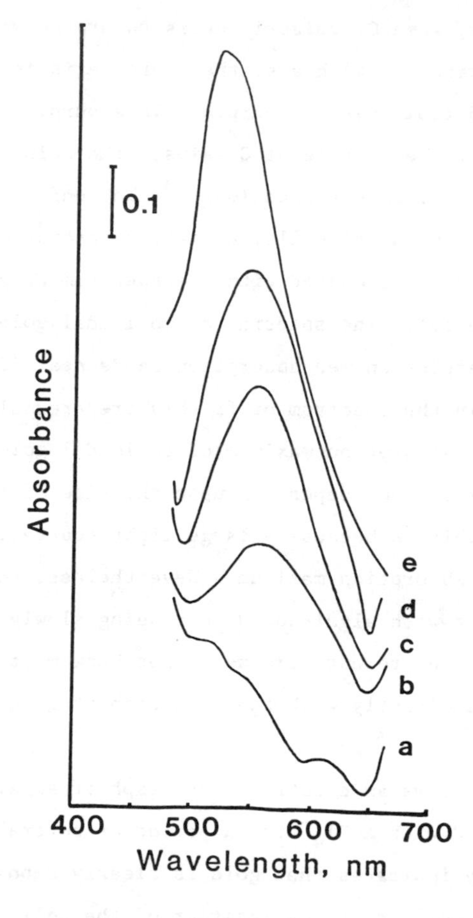

Figure 10. Spectrophotometric Determination of Gold(0) on Chlorella vulgaris. The formation of gold(0) (λ_{max}, 525 nm) on tetrachloroaurate(III)-saturated Chlorella vulgaris was monitored by spectrophotometric analysis of algal suspensions. Algal samples were prepared as follows: Washed C. vulgaris was suspended at 2.0 mg/ml in 0.5 mM hydrogen tetrachloroaurate(III) at pH 2.0, and stirred for one hour. The cells were isolated by centrifugation, washed two times at pH 2.0, and aliquots were stored moist in sealed test tubes. Sepectra were recorded immediately (a), and at 20 hr (b), 69 hr (c), 93 hr (d). Cells were resuspended at 2.0 mg/ml in glycerol immediately before spectra were recorded, to avoid settling. The spectrum of colloidal gold(0) is shown in curve e.

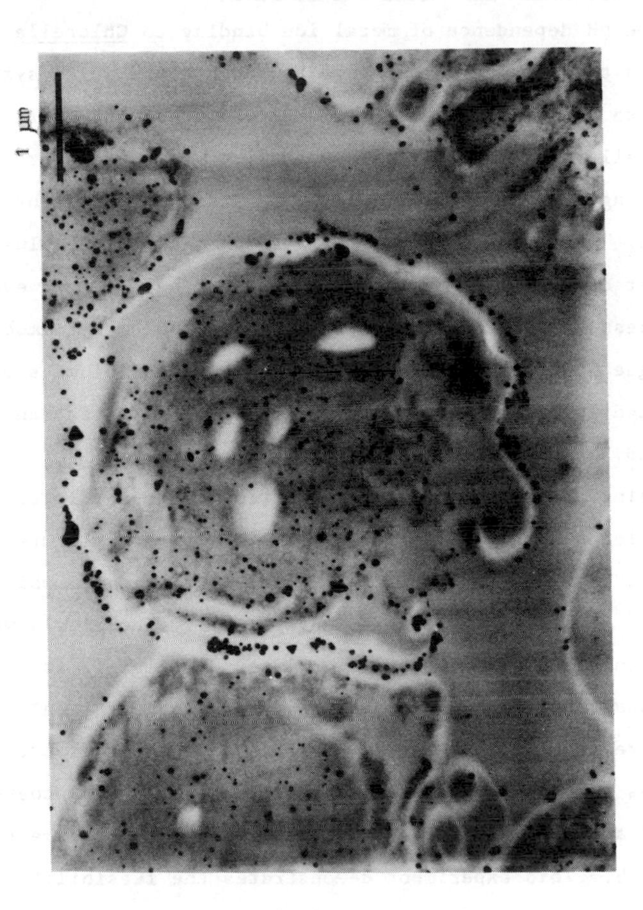

Plate 1. Electron Micrographs of Chlorella vulgaris Which Had Been Previously Exposed to $AuCl_4^-$. The algal samples were prepared by repeatedly exposing an algal pellet to 5 solutions containing 0.5 mM $HAuCl_4$. The sample was fixed with 2.5% glutaraldehyde and viewed unstained.

Immobilized Algae For Metal Ion Recovery.

The pH dependence of metal ion binding to Chlorella suggests that it may be possible to recycle the algal system much like an ion-exchange resin. It would be most useful if the algae cells could be packed into a column so that waters containing metal ions could simply be passed through the columns. Unfortunately when that is done, the algae clump together and significant flow rate can not be achieved even with high pressures. We have alleviated this problem by immobilizing the algae in polymer matrices. The immobilized biomass can then be packed into columns through which high flow rates can be achieved.

Using a combination of pH variation and specific elution we were able to separate the individual ions from a mixture of Zn^{2+}, Cu^{2+}, Hg^{2+}, and $AuCl_4^-$ using polyacrylamide-immobilized algae.[19] This is illustrated in Figure 11. The metals were loaded on a short column of immobilized C. vulgaris at pH 6.0. After washing the column thoroughly at the same pH, Zn^{2+} and Cu^{2+} were sequentially eluted by means of a pH gradient. The Hg^{2+} was then collected by elution with 0.5M 2-mercaptoethanol at pH 2 and gold was collected by elution with the same reagent at pH 5.0. This experiment demonstrates the feasibility of selective recovery of bound metals from immobilized algae.

A method for immobilizing Chlorella in polyacrylamide was described by Nakajima et al.[25,26]. The resulting preparations function satisfactorily under certain conditions but are not very durable. At alga concentrations above 20% (on a dry weight basis), the material is extremely prone to fracture. We have now developed a technique for immobilizing algal cells in silica.[24] This alga-silica is far superior to

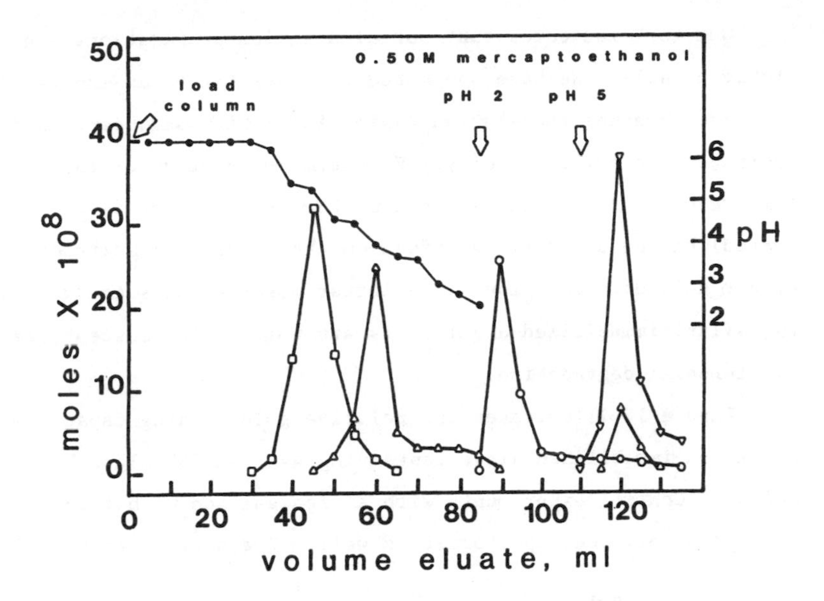

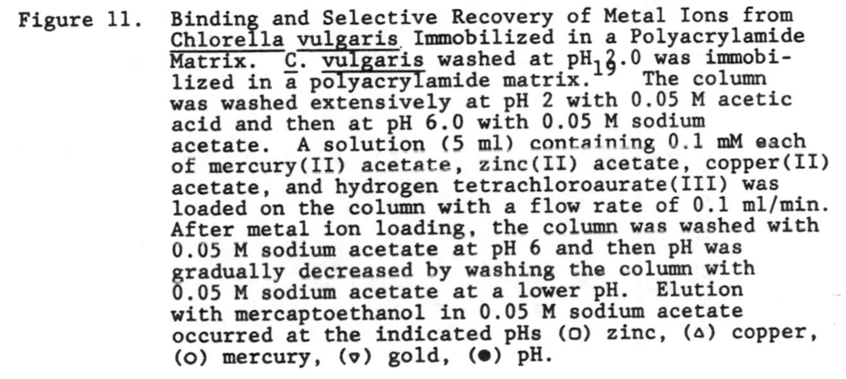

Figure 11. Binding and Selective Recovery of Metal Ions from Chlorella vulgaris Immobilized in a Polyacrylamide Matrix. C. vulgaris washed at pH 2.0 was immobilized in a polyacrylamide matrix.[19] The column was washed extensively at pH 2 with 0.05 M acetic acid and then at pH 6.0 with 0.05 M sodium acetate. A solution (5 ml) containing 0.1 mM each of mercury(II) acetate, zinc(II) acetate, copper(II) acetate, and hydrogen tetrachloroaurate(III) was loaded on the column with a flow rate of 0.1 ml/min. After metal ion loading, the column was washed with 0.05 M sodium acetate at pH 6 and then pH was gradually decreased by washing the column with 0.05 M sodium acetate at a lower pH. Elution with mercaptoethanol in 0.05 M sodium acetate occurred at the indicated pHs (□) zinc, (△) copper, (○) mercury, (▽) gold, (●) pH.

polyacrylamide-embedded material. It is extremely hard ("rocklike") and resists fragmentation. Nevertheless, it is sufficiently porous that all potential metal-ion binding sites are capable of being occupied. Moreover, the algal content of the polymer can be made a high as 90% (on a dry weight basis) without compromising its structural integrity. The algae-silica material functions superbly as a chromatographic matrix.

Our work indicates that our alga-silica preparations are highly durable. We have subjected the material to as many as 30 cycles of binding and elution (using Au^{3+}) without observing any decrease in binding capacity. Furthermore, storage at room temperature for as long as three months (the longest period tested) has no deleterious effect on the binding capacity of either gold or copper ion. The latter observation suggests that the silica-immobilized algal cells are not readily susceptible to microbial degradation.

Figure 12 illustrates the relative gold binding capacities of silica-immobilized algae containing 44% and 86% algae by dry weight. Control experiments with silica gel and with free Chlorella cells were performed as well. The data in Figure 12

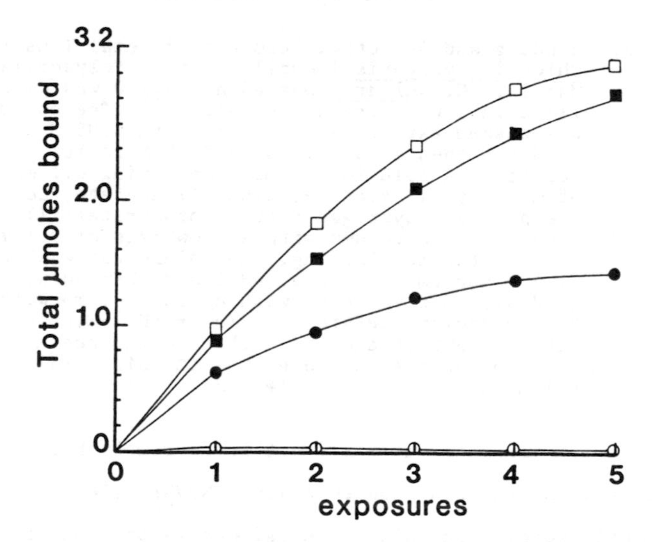

Figure 12. Relative Gold Binding Capacities of Algae, Algae Immobilized in Silica Gel, and Silica Gel. Silica gel (○), 44% algae-silica gel (●), 86% algae-silica gel (■), and algae (□) were suspended at 2.0 mg/ml in 5.0 ml of 0.2 mM $AuCl_4^-$ for a total of five exposures.

indicates that as the algal mass in the polymer increases, so does the gold binding capacity. The pores of the polymer are apparently large enough to allow free diffusion of ions since similar quantities of metals are bound by free and immobilized cells.

Figure 13 shows results of experiments with an electroplating waste solution. This waste solution was an ammoniacal copper solution, pH 8.3, obtained from a company in the printed circuit industry. The waste solution was passed through a algae-silica column without pretreatment. The initial effluents showed the absence of copper. Once the column effluents showed breakthrough of copper, the bound copper was eluted with sulfuric acid.

CONCLUSIONS

The algae-silica matrix functions as a "biological," mixed-bed ion-exchange resin. Like ion-exchange resins, the algae-silica materail can be recycled. We have sorbed and stripped metal ions over as many as 30 cycles with no noticeable loss in efficiency. In contrast to some ion exchange resins, however, a real advantage of the algae-silica matrix is that the compounds of hard water (Ca^{+2}, Mg^{+2}) or monovalent cations (Na^{+}, K^{+}) do not significantly interfere with the binding of toxic, heavy metal ions. The binding of Ca^{+2} and Mg^{+2} to ion-exchange resins often limits ion-exchange usefulness since these ions are frequently present in high concentrations and compete for heavy metal ion binding. This means that frequent regeneration of ion-exchange resins is necessary in order to remove heavy metal ions from solutions.

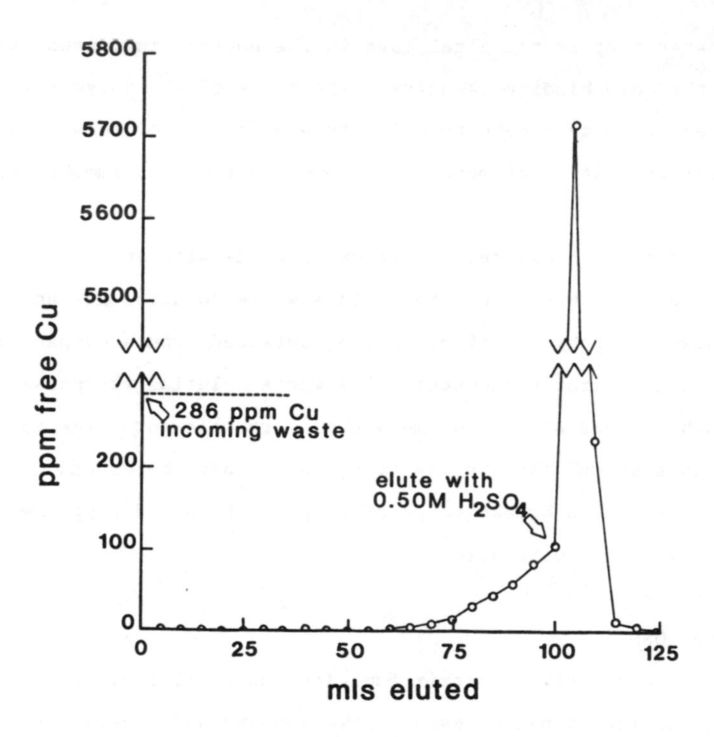

Figure 13. Removal and Recovery of Copper from an Industrial Waste Water Solution. A waste solution containing copper as the ammonia complex was passed through a column containing 0.5 g of silica-immobilized algae. After the column became saturated as evidenced by the presence of copper in the effluent (near 75 ml), sulfuric acid was passed over the column to elute the bound copper ion.

Another advantage of the algae-silica system is that it can be used, under certain conditions, to remove a variety of heavy metals from solutions, and then these metal ions can be selectively stripped from the matrix. Because certain metal ions are bound with different affinities depending upon the pH of the medium only a pH gradient is needed in the regeneration cycle in order to separate metal ions.

An additional advantage of the algae-silica system is that the material costs should be more than competitive with those associated with an ion-exchange system. Thus the use of the algae-silica technology for wastewater treatment and for treatment of mining process streams appears to offer an attractive alternative for various industries.

REFERENCES

[1] D. Khummongkol, G.S. Canterford, and C. Fryer, Biotechnol. Bioeng., 1982, 24, 2643.

[2] M. Galun, P. Keller, D. Malki, H. Feldstein, E. Galun, S.M. Siegel, and B.Z. Seigel, Science, 1983, 219, 285.

[3] M. Galun, P. Keller, D. Malki, H. Feldstein, E. Galun, S.M. Siegel, and B.Z. Seigel, Water, Air and Soil Pollut., 1983, 20, 221.

[4] A. Preston, D.F. Jefferies, J.W.R. Dutton, B.R. Harvey and A.K. Steele, Environ. Pollut., 1972, 3, 69.

[5] D.R. Trollope and B. Evans, Environ. Pollut., 1976, 11, 106.

[6] F. Laube, S.J. Ramamoorthy and D.J. Kushner, Bull. Environ. Contam. 1979, 21, 763.

[7] A. Les and R.W. Walker, Water, Air and Soil Pollut., 1984, 23, 129.

[8] D.S. Filip, T. Peters, V.D. Adams and E.J. Middlebrooks, Water Res., 1979, 13, 305.

[9] J.C. Jannett, J.M. Hassett and J.E. Smith, Proc. Int. Conf. Managenent and Control of Heavy Metals in the Environment, London, 1979.

[10] N.L. Gale and B.G. Wixon, Proc. Int. Conf. Management and Control of Heavy Metals in the Environment, London, 1979.

[11]T. Horikoshi, A. Nakajima and T. Sakaguchi, Agric. Biol. Chem., 1979, 32, 617.

[12]M. Tsezos, Ph.D. Dissertation, McGill University, Montreal, 1980.

[13]M. Tsezos and B. Volesky, Biotechnol. Bioeng., 1981, 23, 583.

[14]M. Tsezos and B. Volesky, Biotechnol. Bioeng., 1982, 24, 385.

[15]M. Tsezos and B. Volesky, Biotechnol. Bioeng., 1982, 24, 9055.

[16]M. Tsezos and B. Volesky, Biotechnol. Bioeng., 1983, 25, 2025.

[17]J. Ferguson and B. Bubela, Chem. Geol., 1974, 13, 163.

[18]A. Nakajima, T. Horikoshi and T. Sakaguchi, Eur. J. Appl. Microbiol. Biotechnol., 1981, 12, 76.

[19]D.W. Darnall, B. Greene, M. Henzl, M. Hosea, R.A. McPherson, J. Sneddon and M.D. Alexander, Environ. Sci. Technol., 1986, 20, 206.

[20]B. Greene, M. Henzl, M. Hosea and D.W. Darnall, Biotech. Bioeng. 1986, in press.

[21]B. Greene, M. Hosea, R.A. McPherson, M. Henzl, M.D. Alexander and D.W. Darnall, Environ. Sci. Technol., 1986, in press.

[22]M. Hosea, B. Greene, R.A. McPherson, M. Henzl, M.D. Alexander and D.W. Darnall, Inorg. Chim. Acta, 1986, 123, 161.

[23]R.J. Puddephatt, "The Chemistry of Gold" Elsevier Scientific Publishing Co., Amsterdam, 1978.

[24]M. Hosea, B. Greene, R.A. McPherson, M. Henzl, M.D. Alexander and D.W. Darnall, unpublished.

The Use of Gram-positive Bacteria for the Removal of Metals from Aqueous Solution

By I.C. Hancock

DEPARTMENT OF MICROBIOLOGY, THE MEDICAL SCHOOL, UNIVERSITY OF NEWCASTLE UPON TYNE, FRAMLINGTON PLACE, NEWCASTLE UPON TYNE NE2 4HH, UK

The use of bacteria, such as Thiobacillus, in metal leaching operations has received a great deal of study and a good theoretical basis has been established for the role of the microorganisms in the process [1]. Although the use of microorganisms for accumulating metal ions from aqueous salt solutions has also been investigated by many research groups, much of the work has been largely empirical, involving blanket screening of microorganisms for the recovery of specific metals, particularly gold, silver and radionuclides arising from nuclear processing. Thus, microorganisms as disparate as Gram-positive bacteria of the genus *Streptomyces* [2], the Gram-negative bacteria *Pseudomonas* [3],the green alga *Chlorella* [4] and the fungus *Rhizopus* [5] have been reported to have particular advantages for the recovery of uranyl ions.

If microbes are to be used in high technology processes for the recovery of precious metals, the removal of toxic heavy metals, or the purification of materials for processes such as electrolytic metal recovery, we need to understand in far more detail what cell components are involved in metal ion binding, where they are located in the cell, the chemical nature of the complexation and the way it may be controlled or modulated. We are studying these questions using model systems, and Table1 shows the sorts of properties that we believe are required.

TABLE 1-REQUIREMENTS FOR A GOOD MICROBIAL METAL-BINDING SYSTEM

1. Microbes remain viable under operating conditions
2. Microbes have high 'extracellular' binding capacity
3. Binding must be effective over a wide range of pH values
4. Binding must exhibit high selectivity for the metal ion of choice
5. The microorganisms must be cheap to grow and recover
6. Recovery of metal ions must be easy and must avoid lysis of the microbe

Many metal ions are toxic to microorganisms,but for maximum flexibility of use microbes should remain viable during metal accumulation. Thus, microbes that can bind large amounts of metal ions while excluding them from the cell cytoplasm where they exert their toxicity are preferable. In the following sections I will describe the attributes of a group of microorganisms, the Gram-positive bacteria, that fulfil, or can be engineered to fulfil these requirements.

TABLE 2-FAVOURABLE PROPERTIES OF GRAM-POSITIVE BACTERIA

1. Main metal-binding sites are in the cell wall, outside the cell's permeability barrier
2. Cell wall is thick, porous and anionic, with high metal-binding capacity
3. Anionic composition, charge density and porosity of the wall can be controlled
4. Many species are cheap to grow and easy to immobilise

1. Properties of Gram-positive bacteria

Table 2 shows some of the attributes of Gram-positive bacteria that make them particularly suitable for metal ion removal.

While the cell wall of a Gram-positive bacterium is a fairly homogeneous layer consisting of a few major macromolecules covalently linked to the principle shape-maintaining component, peptidoglycan, the Gram-negative envelope is a two-layered structure in which an 'outer membrane' of complex structure, susceptible to damage under mild conditions, surrounds a very thin layer of peptidoglycan that carries no covalently linked polyanions.

The Gram-positive cell wall can be considered as a layer of microporous ion-exchanger[6], having the additional attribute of non-ionic functional groups capable of complexation with transition metal cations. In the bacteria Bacillus subtilis and Bacillus licheniformis , which we and others have used for model studies, the wall is about 30nm thick, may be porous to polyethylene glycols up to a molecular weight of 60,000[7] and to globular proteins up to a molecular weight of 70,000[8] and has an anionic charge density of up to 1 milliequivalent/ml of wet, packed walls or 5 milliequivalents/g dry weight. Studies of surface charge using particle electrophoresis over a range of pH values reveal the presence of anionic groups that titrate in the pH ranges 1.5 to 3 and 3 to 5 and these can be identified as phosphodiester and carboxylate groups respectively. Because of the high porosity of the wall, it exhibits a cation exchange capacity close to its theoretical limit[9] and we have achieved values of greater than 4 mequiv Mg^{2+}/g dry wt wall from appropriate bacteria. The metal-binding behaviour of live bacteria is entirely compatible with the cell wall being the principle site of accumulation [9] and this contrasts with, for example, *Chlorella* [4] where heat-killed cells, in which the permeability barrier to the cell has been disrupted, exhibit much higher metal-binding capacities than live ones. Where binding of metal ions to Gram-positive walls is entirely ionic, apparent association constants of the order of 10^4-10^5 M^{-1} are observed for bivalent cations[10,11] but frequently binding capacities and affinities are higher than can be accounted for by ion exchange processes alone [12,5] and this effect will be discussed later.

2. Cell wall structure.

The basis of the Gram-positive bacterial cell wall is a three-dimensional network of peptidoglycan in which glycan chains of about 100 disaccharide repeating units are cross-linked by short peptide chains of 8 to 12 aminoacids, depending on the species of bacterium. The disaccharide unit consists of two N-acetylated aminosugars, glucosamine and muramic acid, and it is to the 3-O-lactyl ether group of the latter sugar that the peptides are attached. The N-acetylmuramyl residues are also the points of attachment for the 'accessory' polymers teichoic acids and teichuronic acids, which are covalently linked through phosphodiesters to C6 of a proportion of the aminosugar residues. Neutral polysaccharides sometimes occur, linked to peptidoglycan in the same way. In many strains of bacteria these components constitute over 90% of the weight of the wall which is, therefore, essentially a single, giant macromolecule of which about 50% is peptidoglycan. When cell walls are purified from broken bacteria they retain the shape of the cell, but despite many physical studies no clear picture of their molecular architecture has yet emerged. Although the glycosidic linkages between N-acetylaminosugars in the glycan strands are all ß1-4 , the conformation of the polymer backbone does not appear to be analogous to that of cellulose and chitin[13]. Calculations from physical measurements indicate that the amount of peptidoglycan found in the *Bacillus* wall would occupy about 14 monomolecular layers if it was orientated in the plane of the cylindrical wall. The teichoic acid and teichuronic acid chains, which are short polymers of about 40 repeating units, are carried by about 8% of the peptidoglycan repeating units and are distributed throughout the thickness of the wall. These two accessory polymers provide most of the anionic groups of the cell wall and are largely responsible for its ion-exchange properties[9,11]

Figure 1 shows an idealised section of the wall of Bacillus licheniformis, indicating the various charged and polar groups that might be involved in metal complexation. Although the phosphodiesters of teichoic acid and the carboxylate groups in teichuronic acid contribute most of the charge, smaller numbers of carboxylate groups and amino groups may be provided by peptide chains of peptidoglycan that are not cross-linked; cross-linking is rarely complete, and may vary with growth conditions[14,15] Amino groups are also provided by variable amounts of the aminoacid D-alanine that is found esterified to alditol residues in most teichoic acids; in a few species some of the aminosugars in peptidoglycan lack N-acetyl groups, so providing additional cationic groups.

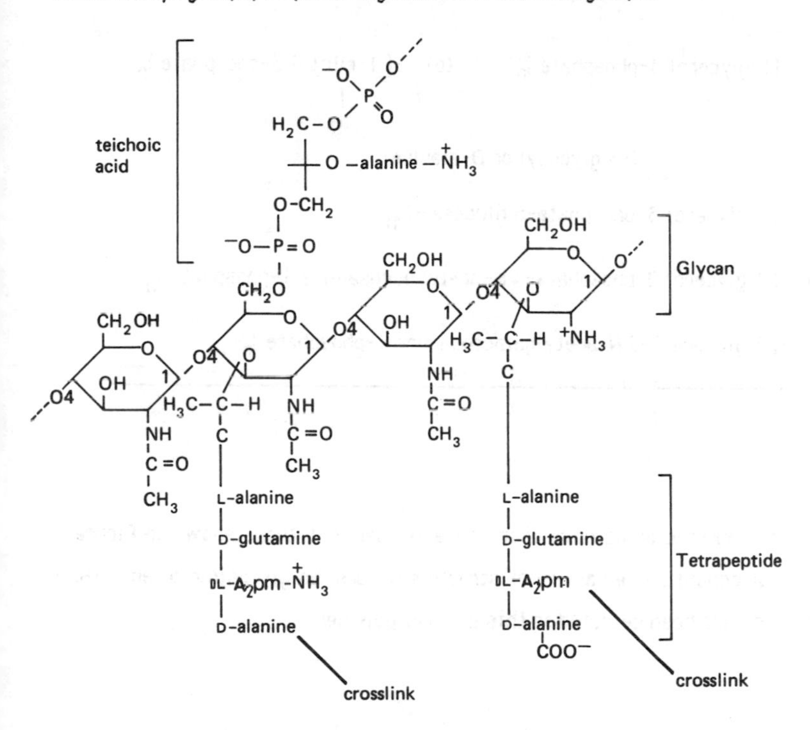

FIGURE 1 - CHARGED GROUPS IN THE CELL WALL

Figure 2 indicates the variety of teichoic acids that have been characterised. The simplest types are polymers of a single species of alditol phosphate - glycerol, ribitol or, rarely, mannitol phosphate - to which monosaccharide residues may be glycosidically linked and alanine residues esterified. Another common type includes a sugar, or a sugar phosphate, as part of the main chain. In a few cases the polymer consists entirely of phosphodiester-linked N-acetylaminosugars[16]. Up to 70% of the alditol residues may carry an alanyl group.

FIGURE 2 - STRUCTURES OF TEICHOIC ACIDS FROM BACTERIAL CELL WALLS

(a) $-[\text{1-glycerol 3-phosphate}]_n-$ (with R attached at glycerol)

(b) $-[\text{1-ribitol 5-phosphate}]_n-$ (with R attached at ribitol)

(R = glycosyl or D-alanine)

(c) $-[\text{1 glycerol 3-phosphate-6 glucose -1}]_n-$

(d) $-[\text{1 glycerol 3-phosphate-4 N-acetylglucosamine 1-phosphate}]_n-$

(e) $-[\text{6 glucose 1-3 N-acetylgalactosamine 1-phosphate}]_n-$

Fewer teichuronic acids have been characterised, but those shown in Figure 3 indicate that considerable variety of structure occurs. No glycosidic or aminoacyl substituents have been detected in this class of polymer.

FIGURE 3-TEICHURONIC ACIDS FROM BACTERIAL CELL WALLS

Bacillus licheniformis

-[4 GlcA ß1-4 GlcA ß1-3 GalNAc ß1-6 GalNAc 1]$_n$-

Bacillus megaterium

-[4 Glc 1-3 Rham 1-4 Rham 1]$_n$-
|
GlcA

Micrococcus luteus

-[4 ManNAcA ß1-6 Glc 1]$_n$-

Under normal laboratory growth conditions, most bacteria contain only teichoic acid or teichuronic acid but occasionally, as in *Bacillus licheniformis*, the two occur together. However, many strains contain more than one type of teichoic acid; the types sometimes differ only in the degree of glycosylation, but some strains contain mixtures of glycerol and ribitol teichoic acids.

3. Metal binding to isolated teichoic acid and teichuronic acids.

The first suggestion that teichoic acids might participate in ion exchange in the bacterial cell wall was made by Archibald et al.[17] in 1961 although cation binding by walls was not demonstrated until 1967[18]. Meers and Tempest[19] observed a correlation between wall cation-binding capacity and teichoic acid content and Heptinstall et al.[9] unambiguously confirmed the relationship by specific chemical removal of teichoic acid by periodate oxidation of walls of *Staphylococcus aureus* that contained a ribitol teichoic acid. The latter work also demonstrated that the presence of positively charged D-alanine residues esterified to the teichoic acid reduced the amount of bivalent metal cation bound by the cell wall under given conditions.

Magnesium ion binding by isolated teichoic acids has been studied in some detail. Teichoic acids were extracted from purified cell walls in dilute acid, which hydrolyses the sugar 1-phosphate linkage that attaches the teichoic acid to C6 of muramic acid residues in peptidoglycan[20], and were exhaustively purified by gel permeation and ion-exchange chromatography. Binding data were derived from equilibrium dialysis experiments using constant weights of teichoic acid at varying total Mg^{2+} concentrations[10,21,22], and were analysed by the treatment of Scatchard[23] which considers ion binding in terms of multiple equilibria.

Figure 4 shows competition between Na^{+},Ca^{2+} and Mg^{2+} for binding sites on a polyglycerol phosphate teichoic acid that carried glucosyl substituents on 50% of the glycerol residues and alanyl ester residues on 21% of the glycerols. The experiments were carried out at pH 5 to ensure stability of the esterified alanine. Within the range of the experiment, Scatchard plots of the data were linear, indicating equivalence of binding sites and lack of interaction between them. The

FIGURE 4 - SCATCHARD ANALYSIS OF BINDING OF Mg^{2+} TO TEICHOIC ACID

r is the number of Mg^{2+} bound per phosphate group and A is the concentration (mM) of free Mg^{2+} at equilibrium.

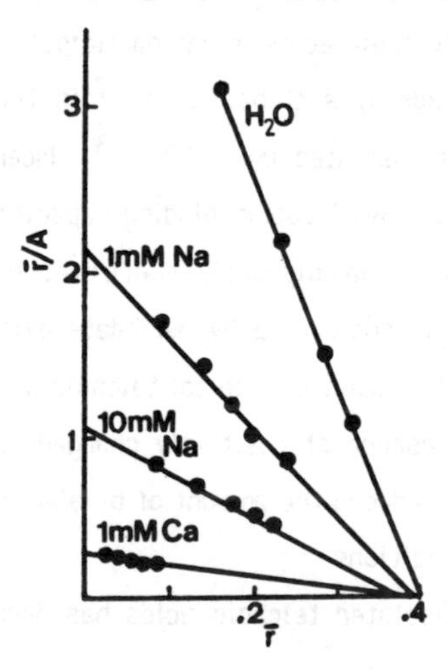

apparent association constant for Mg^{2+}, obtained from the slope of the line, was $1.25 \times 10^4 M^{-1}$ and the figure for Na^+ was about one order of magnitude lower. The total number of binding sites, given by the intercept of the line on the r' axis, was 0.42 equivs/mol phosphate. If it can be assumed that bound alanyl groups neutralised 20% of the phosphate negative charges, this figure indicates potential binding of 1 Mg^{2+} for every two available phosphate groups.

TABLE 3 - BINDING OF MAGNESIUM IONS TO TEICHOIC ACIDS and TEICHURONIC ACIDS

polyanion	bacterial source	Mg/anion *	K_a (M^{-1})
glucosylated polyGP (P:Glc:Ala 1: 0.5 : 0.21)	Lactobacillus buchneri	0.42	2.7×10^3
as above, but P:Glc:Ala 1 : 0.5 : 0.02		0.51	2.9×10^3
N-acetylglucosaminyl polyRP (P:GlcNAc:Ala 1 : 1 : 0.67)	Staphylococcus aureus	0.34	0.83×10^3
as above, but P:GlcNAc:Ala 1 : 1 : 0.04		1.0	0.63×10^3
teichuronic acid (pH 7) (GlcA : GalNAc 1:1)	Bacillus subtilis	1.0	0.3×10^3

* Unless otherwise stated, measurements were made at 10mM Na^+, pH 5.0, with 1mM phosphate or carboxylate in the polyanion.

Comparison of the behaviour of different wall polyanions with respect to magnesium ions revealed some intriguing differences (Table 3). Where internal neutralisation of anionic sites by alanine was not possible, the stoichiometry of bivalent cation binding depended on the structure of the polyanion. While poly(glycerolphosphate) bound magnesium ions with the expected stoichiometry of one Mg^{2+} for every two polymer phosphodiester groups, poly(ribitolphosphate) and the teichuronic acid bound, with less avidity, one bivalent cation to each monovalent anionic site. In these latter cases the second positive charge on the cation must have been satisfied by a mobile anion in solution. Space-filling models of the teichoic acids show quite clearly that while the poly(glycerolphosphate) type of teichoic acid can readily take up a conformation in which Mg^{2+} is bound between two adjacent phosphodiester groups within a polymer chain, poly(ribitolphosphate) can not adopt such a conformation without considerable strain. Moreover, such a folding process would be severely sterically hindered by the bulky glycosyl substituents on the ribitol residues.

Doyle et al.[24] showed, by sedimentation and viscosity studies that poly(glycerolphosphate) adopted different conformations in the presence of magnesium and sodium salts. The presence of magnesium ions supported the same extended, rigid rod structure that the polymer adopted at low ionic strengths, whereas in sodium salts the polymer assumed a flexible random coil conformation. Thus, the effect of magnesium ions was consistent with intramolecular salt bridge formation, whereas sodium ions simply reduced charge repulsion within the polymer chain, permitting more flexibility in the chain. Unfortunately no comparable information for poly(ribitolphosphate) teichoic acids is available. In the case of teichuronic acid[26], an increase in the concentration of magnesium ions from 1 to 10mM led to an increase in sedimentation coefficient from 1.1 to 1.4S, which was interpreted as evidence for intermolecular cross-linking by the bivalent cations . We examined the mode of binding of magnesium ions to poly(glycerolphosphate) using X-ray photoelectron spectroscopy (ESCA) to measure the binding energies of the magnesium outer shell (2s) electrons and thus obtain information about the effect of binding to teichoic acid and other ligands on the magnesium ion[26]. By this technique it was possible to distinguish between Mg^{2+}

in $MgCl_2$, where the cation is associated with two monovalent anions, and in $MgHPO_4.3H_2O$, where it is neutralised by bivalent phosphate groups. The binding energy of the magnesium 2s electrons within the cation was measurably lower in the latter case. The 2s binding energy of Mg^{2+} bound to poly(glycerolphosphate) resembled that of cations in magnesium hydrogen phosphate, but was higher, approaching that of magnesium in magnesium chloride, if the poly(glycerolphosphate) was heavily esterified with D-alanine (Table 4).

TABLE 4 - MAGNESIUM 2S BINDING ENERGIES

Preparation	Binding energy (eV)
Magnesium chloride	90.7
Magnesium phosphate	89.6
Teichoic acid*	90.6, 89.4
Teichoic acid with alanine removed	89.2

* Wall teichoic acid from Lactobacillus buchneri

In conjunction with the other data, we can interpret this in terms of the different modes of magnesium ion binding shown in Figure 5. The metal cation can bind strongly to two adjacent teichoic acid phosphate groups, but when such a configuration is prevented for steric reasons or, as in this case, by internal neutralisation of some of the phosphate negative charges by the amino groups of esterified alanine residues, weaker binding involving a single phosphate and a mobile monovalent cation occurs. Almost identical identical results with dried samples of the wall teichoic acid of Lactobacillus plantarum, which is a glycosylated poly(ribitolphosphate) , indicated that where the polymer chains are close together, a magnesium ion can be bound strongly between phosphates in separate chains. Such a situation may occur in the cell wall, and is discussed in the next section.

FIGURE 5 - ALTERNATIVE MODES FOR BINDING OF Mg^{2+} TO TEICHOIC ACID

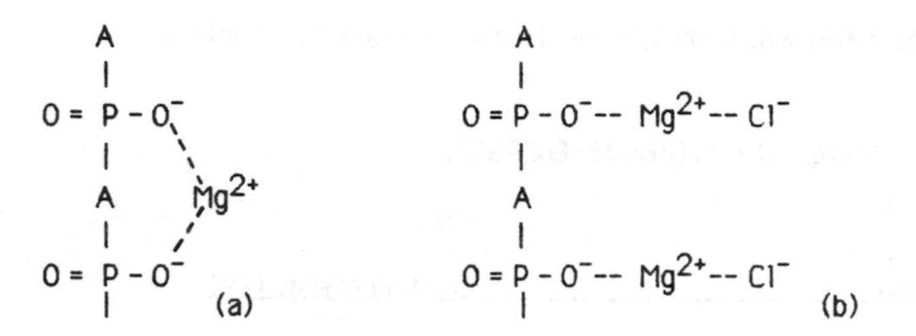

4. Metal ion binding to peptidoglycan.

The peptidoglycan of *Bacillus*, even if fully crosslinked, would contain many free carboxyl groups arising from the dicarboxylic aminoacids, glutamic acid and diaminopimelic acid, in the peptide cross-bridges. In most of the strains studied, however, the carboxyl group of either the glutamic acid or the diaminopimelic acid is amidated[27], so that the net carboxylate negative charge on a fully cross-linked repeating unit is one. The degree of cross-linking in intact walls depends on the strain of bacterium, varying from about 20% in *Micrococcus luteus* to greater than 80% in *Staphylococcus aureus*. In strains of *Bacillus* about 50% of peptidoglycan repeating units are not crosslinked[13], and these units contribute one additional carboxylate anion from the terminal aminoacid of the peptide chain and an amino cation from the subterminal one. It is not known whether there is sufficient flexibility in the peptide to allow the two groups to neutralise one another, and where metal binding by walls stripped of accessory polymers has been studied, contradictory results have been obtained. The peptidoglycan from B.licheniformis had less than 10% of the Mg-binding capacity of intact walls[25] whereas that from B.subtilis had as high a capacity as the native walls, binding almost one order of magnitude more Mg^{2+} than could be accounted for in terms of the fixed anionic charge in the wall[13]. In this case a small amount on ionic binding to the peptidoglycan apparently nucleated the precipitation of unidentified insoluble metal salts, but this was not observed by Doyle et al.[12] using supposedly identical material.

5. Metal binding to intact cell walls

Table 5 compares the binding of Mg^{2+} to cell walls of B.subtilis W23 containing a poly(ribitolphosphate) teichoic acid or a teichuronic acid, with binding to the isolated polyanions. The increased affinity and changed stoichiometry of binding in the cell walls was consistent with a mode of binding in which the bivalent cation bridges between anionic sites on adjacent polymer chains, of the kind observed in dried polyanion samples by ESCA. Consistent with that model, the presence of alanyl residues on the teichoic acid reduced the affinity of the walls for the bivalent cation, as well as reducing the stoichiometry of binding[10].

A further feature of binding to the walls that was not observed with isolated polyanions was negative cooperativity of binding at low cation concentrations. This did not occur with walls in which the teichoic acid was heavily substituted with alanine, but in that case positive cooperativity occurred at high cation concentrations[10], a phenomenon also observed with cell walls containing poly(glycerolphosphate) by Doyle et al.[12]. This suggests that conformational changes may be important, as might be expected given the known changes in polyanion conformation in solution induced by cation binding (see section 3, above) and the changes in wall packing density that occurs in response to changes in the net charge on the wall.

TABLE 5 - MAGNESIUM BINDING BY CELL WALLS

Preparation	Stoichiometry(Mg/anion)	K_a (M^{-1})*
Teichuronic acid	1.0	0.3×10^3
Wall containing teichuronic acid	0.57	8.4×10^3
Teichoic acid	1.0	0.6×10^3
Wall containing teichoic acid	0.52	8.5×10^3

*Measured in the presence of 10mM NaCl, pH5

Marquis[28,29] has used measurements of dextran-impermeable volume and refractive index to show that packing of the wall increases when charge repulsion within the matrix is reduced by increased ionic strength or lower pH. IR spectroscopy has shown that these changes are not accompanied by significant changes in conformation of the peptidoglycan chains[13].

Nitrogen and oxygen ligands in the wall might be expected to contribute to metal ion binding, particularly to complexation of transitional metal ions, and such complexation would be more sensitive to conformational effects than purely ionic binding. Coordination complex formation in Gram-positive cell walls has not been investigated, despite its susceptibility to spectroscopic analysis, but it might be involved in determining the marked, though entirely unpredictable differences in affinity and stoichiometry of binding of different metal cations to cell walls. Few studies of binding specificity have been made[12,13,25] and the results offer no basis for a theoretical treatment, although they indicate quite clearly that considerable selectivity is possible and that the nature of the selectivity varies greatly from strain to strain, depending on cell wall composition.

6. Potential for the manipulation of metal-binding properties

Binding capacity and selectivity of binding may be affected by a variety of wall properties: the diversity of anionic and other polar ligands groups, variations in charge distribution within individual polymers and within the wall as a whole, inter-and intra-molecular charge neutralisation in the polymeric wall constituents, and the porosity and packing of the wall matrix. As indicated by the differences in affinities of Mg^{2+} for ribitol and glycerol teichoic acids, the spacing of charged groups along the polyanionic chain is also important in determining metal binding characteristics. In addition, bacterial morphology may affect secondary accumulation of insoluble metal salts following nucleation by bound cations[5,13,30,31]. All these properties can be controlled by appropriate manipulation of growth conditions or by simple genetic alterations, as described below.

Control of polyanion composition of walls. Several species of *Bacillus* have a regulatory mechanism for diverting phosphate away from cell wall synthesis and essential products such as nucleic acids and phospholipids under conditions of nutrient phosphate limitation[32]. This was first demonstrated by Ellwood & Tempest[33], using continuous chemostat cultures in which the rate of bacterial growth can be made to depend on the equilibrium concentration of a single, limiting nutrient. Under phosphate limitation, *B.subtilis* was found to cease teichoic acid synthesis and replace this polymer in its cell wall by the phosphate-free anionic polymer, teichuronic acid. Archibald et al.[34,35] subsequently showed that intermediate stages, in which walls contained predictable mixtures of teichoic acid and teichuronic acid, could be obtained at slightly higher phosphate concentrations. Such walls contain a roughly constant total number of negatively charged groups, made up of different proportions of phosphate and carboxylate. The enzymic basis of this process is understood[36] and mutants have been obtained which are defective in key enzymes of one or other of the biosynthetic pathways (see for example [37]), permitting the manipulation, by variations in nutrient phosphate, of the total negative charge in the wall.

Control of intramolecular charge neutralisation. The amount of D-alanine esterified to teichoic acid in the cell wall varies depending on growth conditions. Both high pH[15] and high sodium chloride concentrations[33] in the culture medium favour a low degree of esterification and the proportion of alditol residues substituted can be modulated in a predictable way in batch or chemostat culture.

Control of cell wall porosity. Wall porosity depends largely on two factors: anionic charge density in the wall matrix, affecting charge-repulsion, and the degree of peptide cross-linking in the peptidoglycan network. The control of charge density was mentioned above. Although the biochemical basis for the changes is not understood, culture conditions are know to affect the degree of peptidoglycan cross-linking in quantitatively predictable ways, both in batch culture[14] and in the chemostat . Thus the proportion of peptide side chains involved in cross linking in the walls of Staphylococcus aureus varied from 90% in batch-grown bacteria from a rich medium to 62% in carbon-limited cells grown in the chemostat. These differences in cross-linking were reflected in differences in the wall porosity, as indicated by the dextran-impermeable volumes of the isolated walls.

Control of cell morphology. The shape of the bacterial cell and the separation of bacteria from one another after cell division are controlled by the assembly of the cell wall[36,38], a process that is particularly sensitive to changes in the rate of synthesis and chemical nature of the wall polyanions. Thus, mutations and changes in growth conditions that affect the synthesis of teichoic acid and teichuronic acid lead to changes in wall growth that can give rise to spherical cells or long filaments of cells in place of the normal short rod shape[37,38] . Similar effects are produced by mutations in the "autolytic" enzymes that are involved in cell separation and wall remodelling during growth. In these mutants ionic wall polymers are not affected. Morphologically altered bacteria have significant advantages for some applications.

7. Conclusions

I have attempted to show that investigations of the processes involved in metal binding to microorganisms can give information of value in making rational choices of microorganisms for particular purposes. In the long term, detailed studies of the individual factors involved will permit the "tailoring" of a bacterium for a particular metal-binding process and will provide a more refined approach than empirical screening of microbes. Gram-positive bacteria appear to be particularly attractive candidates for industrial use because of their ease of growth, their high metal binding capacities and the ease with which their wall properties can be controlled.

References.

[1]D.P.Kelly, P.R.Norris and C.L.Brierley, SGM Symposium, 1979, 29, 263

[2]T.Horikoshi, A.Nakajima and T. Sakaguchi, Eur.J.Appl.Microbiol.Biotech., 1981, 12, 90

[3]G.W.Strandberg, S.E.Shumate, and J.R.Parrott, Appl.Env.Microbiol, 1981, 41, 237

[4]A.Nakajima, T.Horikoshi and T.Sakaguchi, Eur.J.Appl.Microbiol.Biotech., 1981, 12, 76

[5]M.Tsezos and B.Volesky, Biotech.Bioeng., 1982, 24, 385

[6]G.W.Einolf and E.L.Carstenson, Biophys.J., 1973, 13, 8

[7]R.C.Hughes, P.F.Thurman and E.Stokes, Zeit. Immun., 1975, 149, 126

[8]R.Williamson and J.B.Ward, J.Gen.Microbiol, 1981, 125, 325

[9]S.Heptinstall, A.R.Archibald and J.Baddiley, Nature, 1970, 225, 519

[10]P.A.Lambert, I.C.Hancock and J.Baddiley, Biochem.J., 1975, 151, 671

[11]R.J.Doyle, T.H.Matthews and U.N.Streips, J.Bacteriol., 1980, 143, 471

[12] T.J.Beveridge and R.G.E.Murray, J.Bacteriol., 1980, 141, 876

[13] D.Naumann, G.Barnickel, H.Bradaczek, H.Labischinski and P.Giesbrecht,Eur.J.Biochem.1982,125 505

[14] K.H.Schleiffer, W.P.Hammes and O.Kandler, Adv.Microbial Physiol., 1976,13, 246

[15] A.R.Archibald, J.Baddiley and S.Heptinstall, Biochim.Biophys.Acta, 1973, 291, 629

[16] I.C.Hancock and J.Baddiley, in "The Enzymes of Biological Membranes" ed. A.N.Martonosi, Plenum Press, New York, 1985, Vol 2, 279

[17] A.R.Archibald, J.J.Armstrong, J.Baddiley and J.B.Hay, Nature, 1961, 191, 570

[18] C.Cutinelli and F.Galdiero, J.Bacteriol., 1967,93, 2022

[19] J.L.Meers and D.W.Tempest, Soc.Gen.Microbiol.Proc., 1968, 10, P15

[20] J.Coley, A.R.Archibald and J.Baddiley,FEBS Letts., 1977,80, 405

[21] P.A.Lambert, I.C.Hancock and J.Baddiley, Biochem.J., 1975, 149, 519

[22] J.E.Heckels, P.A.Lambert and J.Baddiley, Biochem.J., 1977, 162, 359

[23] G.Scatchard, Ann.N.Y.Acad.Sci., 1949, 51, 660

[24] R.J.Doyle, M.L.McDannell, U.N.Streips, D.C.Birdsell and F.E.Young, J.Bacteriol., 1974, 118,606

[25] T.J.Beveridge, C.W.Forsberg and R.J.Doyle, J.Bacteriol., 1982, 150, 1438

[26] J.Baddiley, I.C.Hancock and P.M.A.Sherwood, Nature, 1973,243, 43

[27] K.H.Schleiffer and O.Kandler, Bacteriol.Rev., 1972, 36, 407

[28] R.E.Marquis, J.Bacteriol., 1968, 95, 775

[29] R.E.Marquis, J.Bacteriol., 1973, 116, 1273

[30] J.C.Jennett, E.Bolter, N.Gale, W.Tranter and M.Hardie, International Symposium on Minerals and the Environment , 1975, The Institute of Mining, London

[31] J.Degens, Nature, 1981, 298, 262

[32] D.C.Ellwood and D.W.Tempest, Biochem.J., 1969, 111, 1

[33] D.C.Ellwood and D.W.Tempest, Adv.Microbial Physiol., 1972, 7, 83-117

[34] A.J.Anderson, R.S.Green and A.R.Archibald, FEMS Microbiol.Lett., 1978, 4, 129

[35] W.K.Lang, K.Glassey and A.R.Archibald, J.Bacteriol., 1982, 151, 367

[36] I.C.Hancock, Biochem. Soc.Trans. 1985, 13, 994

[37] R.L.Robson and J.Baddiley, J.Bacteriol., 1977, 129, 1051

[38] C.W.Forsberg, P.B.Wyrick, J.B.Ward and H.J.Rogers, J.Bacteriol., 1973, 113, 969

Metal Absorption by Modified Chitins

By R.A.A. Muzzarelli,[*,1] and R. Rocchetti[2]

[1]FACULTY OF MEDICINE, UNIVERSITY OF ANCONA, I-60100 ANCONA, ITALY

[2]'V. VOLTERRA' TECHNICAL INSTITUTE, I-60020, ANCONA, ITALY

Chitosan, [(1-4) 2-amino-2-deoxy- β -D-glucan] is a natural chelating polymer, obtained by deacetylation of chitin, which collects transition metal ions thanks to its hydroxyl and amino groups, high hydrophilicity and specific interactions with oxyanions. Chitosan, having a degree of deacetylation around 58 %, has been used to measure the uptake of metal ions from solutions and found to possess remarkable capacity, which is of course pH and concentration dependent. Capacity, in fact, decreases with decreasing concentration of metal ion in solution [1, 2].

The chelating ability of chitosan for metal ion collection from solutions and waters, including sea water, is well documented [3 - 8] : the same can be said about its specificity, because transition metal ions are preferentially collected from brines. Of particular significance are the data obtained on sea water: both the chelating resin Dowex A-100 and chitosan reportedly collect only a small part of the total copper present in untreated sea water [9]. Evidently, metals occur in complex forms which are more or less retained by the polymers, because of their stability [10]. Complete metal recovery is possible only when oxidative destruction of organic matter is carried out with persulfate as a preliminary step. In fact, metals occur in sea water in the following three chemical forms: 1) free hydrated ions; 2) ions forming weak complexes less stable than the chitosan metal chelates; 3) ions strongly complexed by natural soluble substances. The latter are not retained by chitosan columns, while forms 1 and 2 are collected by chitosan from untreated sea water. It was demostrated that Cd occurs in the first two forms [2]. Other elements however, are found in a mixture of the three forms indicated. Whereas the complete collection of Cd from sea water does not require oxidative treatment the completè collection of the other elements is only achieved after the oxidative destruction of organic substances. The experimental data (Table I) show that chitosan collects Cd quantitatively from untreated sea water, while in case of other ions, natural substances occuring in sea water compete with chitosan and depress the metal uptake on this polymer.

The chemical derivatization of chitosan offers the possibility of introducing effective chelating groups in the polymer, thus enhancing its capacity, and making its physical form more suitable, while keeping its favorable characteristics, especially the nature of a hydrophylic polymer and a polyamine.

We have recently reported on the preparation and characterization of novel semi-synthetic polysaccharides obtained from chitosan, particularly aspartate glucan, glycine glucan and serine glucan [11, 12]. They belong to a class of compounds that can be prepared by reacting aldehydo acids or keto acids with the amino groups of chitosan [13 - 15]. These products possess a β(1-4) polysaccharide backbone that carries aminoacid units linked via their nitrogen atoms to the carbons 2 of the anhydroglucose rings. The degree of substitution can be modulated according to the stoichiometry preferred, the degree of deacetylation of the starting chitosan and molecular sizes. Their chemical functions are the carboxyl groups, the secondary amine, the

TABLE I

Mean concentrations of metals (µg/l + standard deviation) for single sea water samples, passed through a 100-mg chitosan column eluted with sulfuric acid.

Metals	After oxidation µg /l (a)	Untreated sea water µg/l (b)	Collection yield % (c)	Untreated sea water + addition µg/l (d)	Addition recovery % (e)	After oxidation + addition µg/l (f)	Addition recovery % (g)
Cadmium	0.16±0.03	0.16±0.02	100	1.15±0.07	99	1.17±0.33	101
Lead	0.20±0.05	0.13±0.03	65	0.83±0.08	70	1.22±0.07	102
Nickel	1.54±0.19	0.94±0.16	61	2.14±0.54	60	3.44±0.45	95
Copper	5.13±0.75	2.64±0.19	52	3.60±0.42	48	7.05±0.41	96

(c), obtained from (b) x 100 /(a)

(e), obtained from (d) - (b) x 100/addition

(g), obtained from (f) - (a) x 100/addition

primary and secondary alcohols in addition to the functions carried by the aminoacid side chains. We have also reported on the novel derivative of chitosan similarly obtained from dehydroascorbic acid [16, 17]. These polymers are finding uses in the medical field [18, 19]; in the nuclear field, for the removal of radionuclides from waste waters [20], and in biotechnology, for the entrapment of cells [21].

While a number of chelating resins can be used for the removal of transition metal ions from waters, the chitosan-derived chelating agents may find use in certain fields where their hydrophilicity and high capacity at low metal concentrations represent advantages over synthetic chelating resins. Because in the preparation of N-substituted chitosans there is interest in having a high proportion of amino groups in free form, fully deacetylated chitosan was prefered.

The derivatives of fully deacetylated chitosan so far studied from the standpoint of the chelating ability are those listed in Table II. They include nine aminoacid glucans [20] and two sugar acid glucans [21 - 23] obtained by reacting chitosan from Euphausia superba (Antarctic krill) having a residual acetylation degree of less than 3 %.

Preparation of the aminoacid glucans.

Chitosan (10 g) with pK value 6.2 and acetylation degree 3 % was suspended in water (1.5 l) and an excess (1.5 mol per mol glucosamine) of ketoacid was added (Table II). At this stage, green and black colors were observed for β-hydroxypyruvic acid and for p-hydroxyphenylpyruvic acid, respectively. The resulting pH value (around 3) was slowly adjusted to 4.5 with 0.2 M NaOH. The insoluble product was then reduced with sodium cyanoborohydride dissolved in water (5.85 g in 50 ml) and the pH value was finally adjusted to about 7 with 1 M NaOH. After 48 h the polysaccharides in the aminoacid form were washed, dialyzed against water and isolated. They were extracted in Soxhlet

Table II. Aminoacid and sugar acid derivatives of chitosan obtained by Schiff reaction and subsequent hydrogenation.

Reagent	Product
Aminoacid glucans	
$CHO.COO^-$, glyoxylate	Glycine glucan
$CH_3CO.COO^-$, pyruvate	Alanine glucan
$CH_2OH.CO.COO^-$, β-hydroxypyruvate	Serine glucan
$C_6H_4OH.CO.COO^-$, p-hydroxyphenyl pyruvate	Tyrosine glucan
$C_6H_5CH_2CO.COO^-$, β-phenylpyruvate	Phenylalanine glucan
$COOH.(CH_2)_2CO.COO^-$, α-ketoglutarate	Glutamate glucan
$CH_3S.(CH_2)_2CO.COO^-$, α- ketomethyolbutyrate	Methionine glucan
$(CH_3)_2CH.CH_2CO.COO^-$, α-ketoisocaproate	Leucine glucan
$COOH.CH_2CO.COO^-$, oxalacetate	Aspartate glucan
Sugar acid glucans	
$CH_2OH.CO.(CHOH)_3.COO^-$, 5-ketogluconate	Gluconate chitosan
$CH_2OH.CHOH.CH.CO.CO.CO$ (ring: CH–O–CO), dehydroascorbate	N-(Dihydroxyethyl) tetrahydrofuryl chitosan

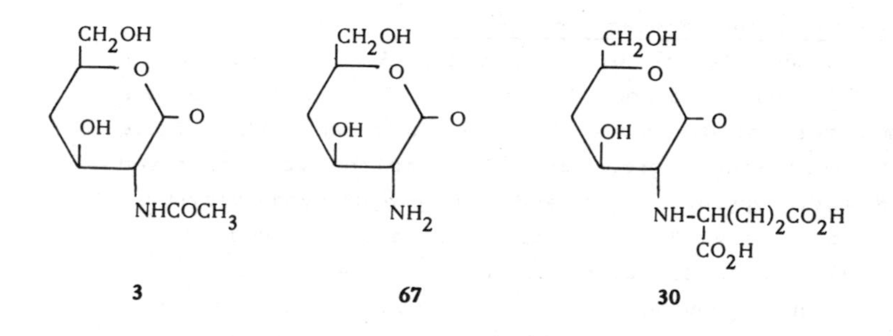

β (1-4) linked units in glutamate glucan (or glutaryl chitosan) and degrees of substitution.

with ethanol, acetone and ether. Their infrared spectra were recorded after conditioning at pH 1.

While not all of the modified polysaccharides listed in Table II possess the same level of chelating ability, some of them are powerful chelating agents.

Chelating capacity.

The EDAX analysis of chitosan and glutamate glucan show that in the course of the preparation, no metallic impurities are introduced in the product, which is found to be free from metals. The EDAX determination of the chelating ability of glutamate glucan for Cr, Co, Ni, Cu and Zn simultaneously present in solution (prior to atomic absorption spectrometry) indicates that the order of affinity is Cu > Cr > Zn > Ni > Co. The competitive chelation is pH-dependent, as indicated in Table III which also shows that sulfate, the anion of salts used for this survey, is collected to a large extent. For another group of metal ions, simultaneously present in solutions (100 ml/l each), EDAX indicates that Hg and Cd are collected to a large extent, whilst Mn is not collected at all. At pH 4 the series is Hg > Cd > Fe > Co > Mn as shown in Table IV.

When the aminoacid glucans were contacted with solutions containing one metal ion (Co 25-250 mg/l, Ni 50-1000 mg/l or Cu 25-250 mg/l) at pH values between 5 and 6, the results shown in Fig. 1 were obtained. They indicate that while Ser-glucan has a lower capacity than Asp-glucan, Gly-glucan and Glu-glucan, it is still between 1 and 20 % by weight for Co at 6-850 mg/l equilibrium concentration. Fig. 2 shows more data for glutamate glucan in contact with zinc, cadmium and copper.

Adsorption of metal ions on glycine glucan.

The cross-linked glycine glucan was contacted with cobalt, nickel, copper and cadmium solutions at pH values between 2.9 and 3.5, then adjusted with sulfuric acid in the batch mode. The initial concentrations were 0.2 and 0.5

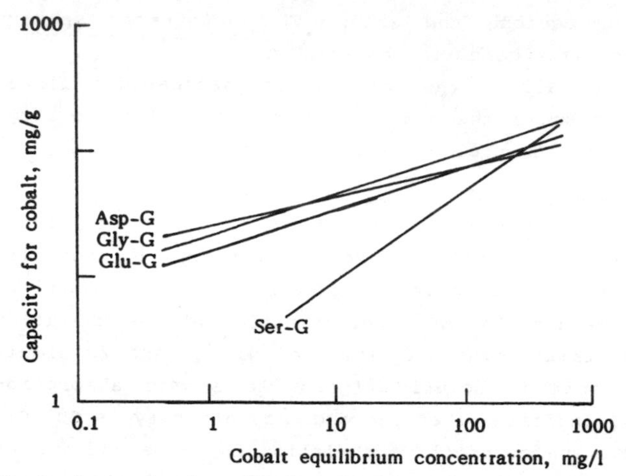

Figure 1. Isotherms for the collection of cobalt on aspartate glucan (Asp-G), glycine glucan (Gly-G), glutamate glucan (Glu-G) and serine glucan (Ser-G), at pH values between 5 and 6.

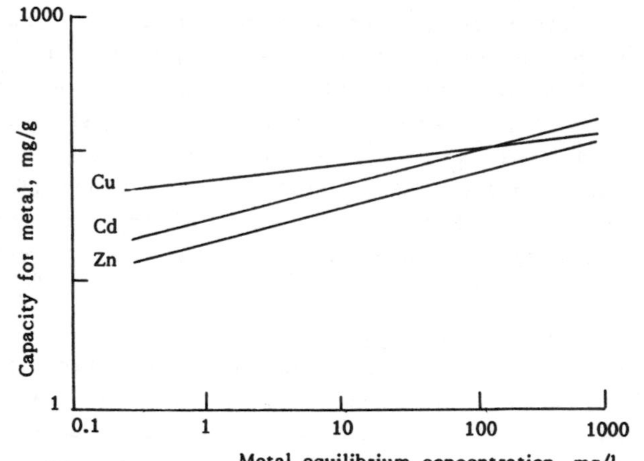

Figure 2. Isotherms for the collection of copper, zinc and cadmium on glutamate glucan, at pH values between 5 and 6.

Table III. Competitive collection of metal ions (as sulphates) on glutamate glucan and aminogluconate glucan, after 1 h contact at 25°C with solutions (25 ml) containing 100 mg/l for each metal. Data are in percent of the total weight of the elements collected.

	pH	Cr	Co	Ni	Cu	Zn	S
Glutamate glucan							
	4.9	15	3	11	22	13	37
	3.9	4	0	3	50	0	43
	3.0	2	0	0	49	0	49
Aminogluconate glucan							
	5.1	17	6	3	19	15	40
	4.1	14	5	1	22	13	45
	3.0	1	0	0	3	0	96

Table IV. Competitive collection of metal ions (all sulphates, except $HgCl_2$) on glutamate glucan and aminogluconate glucan, after 1 h contact at 25°C, with solutions (25 ml) containing 100 mg/l for each metal. Data are in percent of the total weight of the elements collected.

	pH	Cd	Fe	Co	Mn	Hg	S
Glutamate glucan							
	5.9	42	omitted	25	0	15	18
	5.1	26	20	13	0	19	22
	4.0	10	8	6	0	31	45
	3.0	0	6	1	0	24	69
Aminogluconate glucan							
	5.8	27	omitted	1	0	40	32
	5.2	42	20	2	0	0	36
	4.1	7	5	1	0	37	50
	3.1	0	3	0	0	20	77

mM for each metal ion. The collection percentages after 16 h contact were Co, 84.5 - 85.5; Ni, 98.6 - 97.6; Cu, 98.8 - 99.3, Cd, 99.86 - 98.88. The polymers exhibited typical colors, pink, green, blue and yellow, for Co, Ni, Cu and Cd, respectively. Crosslinked glycine glucan was therefore very effective in collecting transition metal ions even from acidic solutions.

<u>Recovery of traces of cobalt and copper from a water by means of crosslinked glycine glucan.</u>

Two chromatographic columns of cross-linked glycine glucan gel were fed with solutions of cobalt and copper, respectively at the concentration of 10 mg/l and at the flow-rate of 0.5 ml/min. The breakthrough points occurred after the passage of 1.8 l of solution. The columns became shorter (4.5 cm) due to the shrinkage of the gel upon chelation, and red and blue colored for Co and Cu respectively. The capacity was found to be about 12 % by weight. A similar experiment was carried out with 1.0 mg/l solutions: the breakthrough points were at 4.0 l and at 12.0 l for Co and Cu respectively, with capacities of 2.0 % and 6.0 % respectively.

Photomicrographs showed that lyophilised glutamate glucan is a spongy substance which offers a large contact surface to fluids. After metal ion collection, the polymers do not reveal any aggregation of metal oxides or growth of inorganic materials, when examined at the SEM. Nevertheless, a sample of glutamate glucan (100 mg) left standing (24 h) unstirred in a Cr(III) solution (100 mg/l) at pH 5, (same conditions as in ref. 12) exhibited scattered small deposits. EDAX analysis for chromium on areas rich in those deposits, however, gave the same results as did clear adjacent areas, thus showing that the proportion of chromium in the deposits is a negligible amount (within instrumental error) compared to the total chromium present in the sample. In other words, the metal ion appears to be nearly homogeneously distributed over the polymer and collected by chelation. No deposit could be detected by SEM on the glutamate glucan, left standing (24 h) in a less concentrated quiescent Cr(III) solution (10 mg/l, pH 5). This finding is in agreement with the high chelating capacity of glutamate glucan compared to chitin/chitosan (see below). It has been stated elsewhere that the chitin surface promotes growth of aggregates which, under certain defined conditions, could represent a significant mode of collection of metal ions. The enhanced chelating ability of glutamate glucan makes chelation to prevent or depress other collection mechanisms (13-15).

Further derivatization of aminoacid glucans.

The N,N'-disubstituted carbodiimides promote condensation between a free amino and a fee carboxyl group to form a peptide link by acid-catalyzed removal of water. Thus an aminoacid glucan, which contains a free carboxyl group, can be coupled with aminoacids containing primary amino groups; the carbodiimide yields an isourea upon hydration. Alcohol groups on the polymer should be protected. To obtain dipeptide glucans, N-carboxymethyl chitosan (1 g) was reacted first with acetic anhydride (20 ml) in the presence of 4-dimethyl amino pyridine (DMAP, 1.1 g) and acetate (0.76 g) to protect the alcohol groups, and the acetylated product, washed with toluene, was combined with either glycine methyl ester, methionin methyl ester or homocysteine thiolactone in the presence of N-ethyl-N'-(3-dimethyl aminopropyl) carbodiimide hydrochloride (DCC). Deprotection of the primary and secondary alcohol groups was then performed with tetrabuthyl ammonium hydroxide [14].

Dipeptide substituted glucans were thus obtained and characterized on the basis of their infrared spectra and elemental analysis. The production of these novel derivatives opens the way to third-generation chitosans carrying oligopeptides particularly suitable for selective and effective chelation of metal ions. It was shown that the tripeptide H-Gly-Gly-His-OH and its N-methyl amide derivative mimic well the Cu(II) and Ni(II) binding site of native human serum albumin[25]. The synthesis and the preliminary conformation properties of a polymeric adduct in which the tripeptide unit Gly-Gly-His has been covalently bound to the ε-amino group of poly(L-lysine) has been recently described. Macromolecular carbohydrate-based analogues of the binding site of human serum albumin become thus easily accessible through peptide synthesis on N-carboxymethyl chitosan.

In vertebrates and fungi, heavy metals are detoxified by the metallothioneins, which are sulfur-rich proteins. Metal chelating compounds have also been proposed for higher plants and a set of novel heavy-metal complexing peptides was isolated from plant cell suspension cultures; the structure of the peptides was established as (γ-glutamic acid-cysteine)$_n$-glycine (n = 3 to 7). These compounds may be viewed as linear polymers of the gamma-Gly-Cys portion of glutathione and may be formed from glutathione itself. They seem to be the simplest (composed of only three different amino acids) natural compounds so far reported that may be engaged in the detoxification and homeostasis of heavy metals through metal-thiolate formation. They lend themselves to coupling to glycine glucan for the purpose of binding heavy metals through mercaptide complexes [26].

Alternative routes to aminoacid glucans.

The feasibility of the grafting of peptides via an amide linkage, onto polysaccharides with free amino groups has been studied [27]. Chitosan reacts easily under heterogeneous conditions with N-carboxyanhydrides of aminoacids, to yield graft copolymers. The heterogeneous conditions favour the applicability of this reaction to chitosan films and beads. The benzyl glutamate and the alanine derivatives were thus obtained. The synthesis of N-tripeptidyl glucosamine by the stepwise reaction of N-carboxy aminoacid anhydride was also reported [28].

Conclusions.

Chitosan, an effective chelating agent, when used as such for the removal of metal ions from natural waters, enters into competition with other natural chelating substances occurring in soluble forms in the waters. Chemical modifications leading to the introduction of an aminoacid function into the glucan backbone, dramatically improves the chelating ability of chitosan, one of the most effective derivatives being the glycine glucan. By taking advantage of the examples offered by natural proteic substances occurring in algae, fungi and animals, the chitosan derivatives can be further modified to yield oligopeptide glucans, where the oligopeptide chains mimic or reproduce the fragment responsible for maximum chelating effect.

In conclusion, it appears that chitosans with varying degrees of deacetylation, are ideal hydrophilic backbones to support aminoacid substituents or oligopeptides linked to C2 through the nitrogen atom. The proper selection of aminoacid sequences would confer the polymer unique and unmatched chelating properties.

Acknowledgement. This work was carried out with the financial contribution of the Italian National Research Council, Progetto Finalizzato Chimica Fine e Secondaria, Roma, Italy (contract No. 84.01211.95).

REFERENCES

1. R.A.A. Muzzarelli, "Chitin", Pergamon Press, Oxford, 1977.
2. R.A.A. Muzzarelli, C. Jeuniaux and G.W. Gooday, "Chitin in Nature and Technology", Plenum Press, New York, 1986.
3. R.A.A. Muzzarelli, "Natural Chelating Polymers", Pergamon Press, Oxford, 1973.
4. T.M. Florence and G.E. Batley, Talanta, 1976, 23, 179.
5. R. Pocklington, Marine Chem., 1977, 5, 479.

6. R.A.A. Muzzarelli, in "The Polysaccharides", G.O. Aspinall, ed., Academic Press, New York, 1985. Vol. 3, Chapter 6.

7. R.A.A. Muzzarelli, in "New Developments in Industrial Polysaccharides" V. Crescenzi, C.M. Dea and S.S. Stivala, eds., Gordon & Breach, New York, 1985.

8. I.F. Slowey and D.W. Hood, Geochim. Cosmochim. Acta, 1971, 35, 121.

9. R.A.A. Muzzarelli and R. Rocchetti, Anal. Chim. Acta, 1974, 69, 35.

10. T.M. Florence and G.E Batley, Talanta, 1975, 22, 201.

11. R.A.A. Muzzarelli, F. Tanfani, M. Emanuelli and L. Bolognini, Biotechnol. Bioengin., 1985, 27, 1115.

12. R.A.A. Muzzarelli, Italian Patent No. 21261, 21 June 1985.

13. R.A.A. Muzzarelli, in "Encyclopedia of Polymer Science and Engineering" A. Mark, B. Bikales, C. Overberger and Z. Menges, eds., John Wiley & Sons, New York, 1985, vol. 3, 2nd edition.

14. R.A.A. Muzzarelli and A. Zattoni, Intl. J. Biol. Macromol. in press.

15. R.A.A. Muzzarelli, F. Tanfani and M. Emanuelli, Carbohydr. Polymers in press.

16. R.A.A. Muzzarelli, F. Tanfani and M. Emanuelli, Carbohydr. Polymers 1984, 4, 137.

17. R.A.A. Muzzarelli, Carbohydr. Polymers , 1985, 5, 85.

18. R.L. Smith, U.S. Patent 4,474,769, 2 Oct. 1984.

19. R.A.A. Muzzarelli, F. Tanfani and M. Emanuelli, Carbohydr. Res., 1982, 107, 199.

20. R.A.A. Muzzarelli and F. Tanfani, Pure Appl. Chem., 1982, 54, 2141.

21. D. Knorr, in "Chitin in Nature and Technology", R.A.A. Muzzarelli, C. Jeuniaux and G.W. Gooday, eds., Plenum Press, New York, 1986.

22. C.A. Eiden, C.A. Jewell and J.P. Wightman, J. Appl. Polymer Sci., 1980, 25, 1587.

23. R. Maruca, B.J. Suder and J.P. Wightman, J. Appl. Polymer Sci., 1982, 27, 4827.

24. B.J. Suder and J.P. Wightman, in "Adsorption from Solutions", R.H. Ottewill, C.H. Rochester and A.L. Smith, eds., Academic Press, New York, 1983, p. 235.

25. M.T. Foffani, S. Mammi, L. Michielin and E. Peggion, Intl. J. Biol. Macromol., 1985, 7, 370.
26. E. Grill, E.L. Wimacker and M.H. Zenk, Science, 1985, 230, 674.
27. S. Aiba, N. Minoura and Y. Fujiwara, Intl. J. Biol. Macromol., 1985, 7, 120.
28. M. Oya and T. Negishi, Bull. Chem. Soc. Japan, 1984, 57, 439.

A New Generation of Solid-state Metal Complexing Materials: Models and Insights Derived from Biological Systems

By I.W. DeVoe* and B.E. Holbein

DEVOE-HOLBEIN TECHNOLOGY BV, ZONWEG 35, 2516 AK DEN HAAG, THE NETHERLANDS

The increasing public awareness of the hazards present in industrial liquid, solid, and gaseous waste emissions puts an ever-increasing pressure on nuclear and non-nuclear industries to conform to safe environmental standards. At the same time, standards for industrial waste discharges are periodically raised as new information is gathered on the effects of such pollutants on public health, on the natural ecology or as more efficient technologies become available. As a consequence, industries are often faced with the dilemma of having to comply with high standards when conventional, cost-effective technology is not available to meet the newly imposed requirements.

Some metal emissions in gaseous and aqueous effluents from nuclear industries pose special problems for which conventional technology provides virtually no solution. Moreover, conventional methods such as precipitation, ion-exchange, reverse osmosis, and electrolysis often do not produce the effluent quality demanded by today's standards. Therefore, the industrialist is faced with the very real situation of having purchased expensive emission control equipment today that can be rendered obsolete tomorrow as a result of newly imposed, more stringent emission standards.

The obvious solution to this dilemma facing industry would be a technology that could remove hazardous metals from water to levels that well exceed standards throughout the world by a comfortable margin. Such technology would prevent future situations in which today's solution will be obsolete tomorrow. Moreover, the best solution should be cost effective and provide a clean effluent with little or no hazardous waste as a by-product.

A new approach toward providing such a solution was sought by DeVoe and Holbein (7) using biological systems as a model for the laboratory chemical synthesis of new metal-capturing compositions. This approach took into account the highly efficient, metal handling systems already employed by living systems.

The living cells of plants, animals and microbes possess extremely efficient mechanisms for the selective use of a variety of metals (e.g. Na, K, Mg, Ca, Cu, Zn, Co, Fe, Se, Mn) that are essential nutrients to sustain life. A cell may acquire a target metal by means of specialized receptor sites on its surface that recognize and bind only one metal species. Such metal-capturing systems usually apply to metals that exist in the soluble state in the range of physiological pH. Examples include the mechanisms for the acquisition of magnesium and phosphorus.

More complex mechanisms exist for the selective capture of some metals, e.g. iron. Iron has a very low solubility in the range of pH where most life forms can survive and grow. In response to the depressed levels of free iron in their environment, most aerobic microbes, for example, synthesize and excrete into their surroundings low molecular weight nonporphyrin, nonprotein iron-binding molecules, collectively known as siderophores. Neilands (14) defines a siderophore as a low-molecular-weight (500 - 1000 daltons), virtually ferric-specific ligand, the biosynthesis of which is to supply iron to the cell. The siderophores generally fall into two classes of molecules, hydroxamates and phenolate-catecholates. Both types exhibit very strong affinity for Fe(III) with formation constants lying in the range of 10^{30} to 10^{50}. Examples of the prototypes of each of these siderophoric classes are shown in Figures 1 and 2.

Prior to work on high-affinity metal capture from water with synthetic materials, DeVoe and Holbein (1,2,4-7,10,11,13,16) were independently involved in research on the pathogenicity of various infectious bacteria with special emphasis on the ability of such virulent microbes to acquire the vital nutrient iron upon entering the body. Iron in the body is never free but can be found complexed to a variety of special molecules such as hemoglobin, cytochromes, and the blood protein transferrin. DeVoe and Holbein had shown that iron deprivation caused bacteria pathogenic in human beings to express gene products giving the microbes strong virulence properties. A constant practical problem in these studies was the difficulty in removing selectively only iron from the *in vitro* growth medium so that the strong virulence features might be expressed outside of the host *in vitro*. To accomplish this DeVoe and Holbein (7) covalently linked the microbial siderophores Desferrioxamine and Enterobactin to solid supports, e.g. porous beads (siderophoric composition, Figure 3).

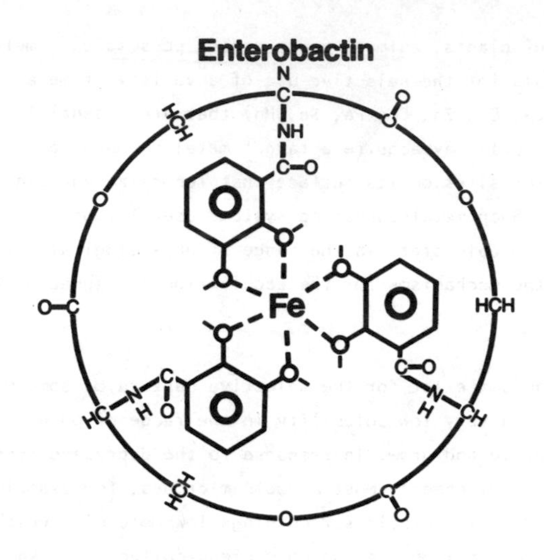

Figure 1. Microbial siderophore Enterobactin

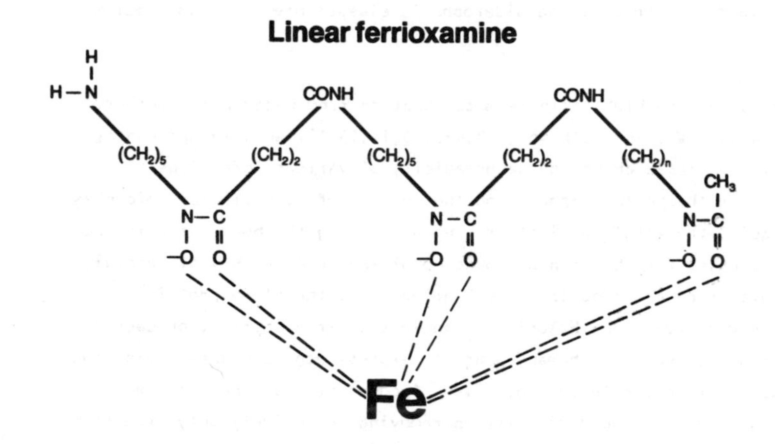

Figure 2. Microbial siderophore Ferrioxamine

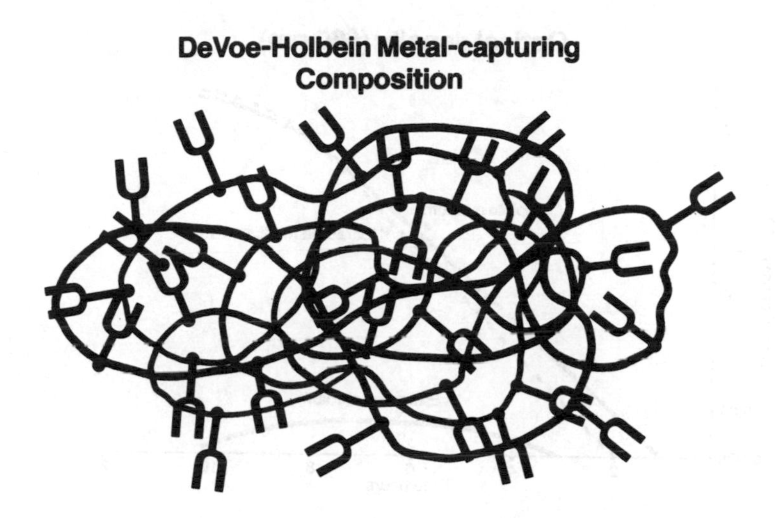

Figure 3. Graphic display of metal-capturing composition. Fork-like symbols represent siderophores immobilized through bridging agents to a solid surface represented by the continuous line.

When these particulate compositions were used as a fixed-bed iron-retrieval system, indeed only iron was removed from liquid growth media. As a consequence of lowering the iron concentration below 0.1μM bacteria were unable to grow (7) in any media so processed unless the iron-free media were once again replenished specifically with this metal (Figure 4). Other media that will support microbial growth, e.g. wine, contact lens rinse solutions, and others, were iron-extracted by the same methods. In each instance, the removal of iron prevented subsequent microbial growth even when the extracted media were inoculated with pathogenic or free-living microbes.

Another feature of these siderophoric compositions, which is particularly relevant currently, is their capability to bind actinides, e.g. plutonium, thorium, uranium, etc, with very high affinity. For example, recently Battelle Pacific Northwest Laboratories in conjunction with West Valley Nuclear Corporation (Westinghouse Electric Co.) tested DeVoe-Holbein siderophoric compositions for their ability to remove selectively trace amounts of plutonium from high-salt residue liquors remaining after nuclear fuel reprocessing (Table 1 and 2).

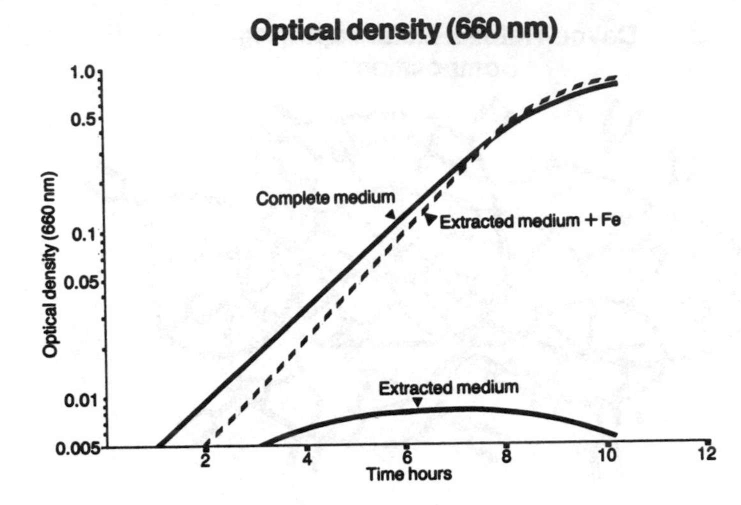

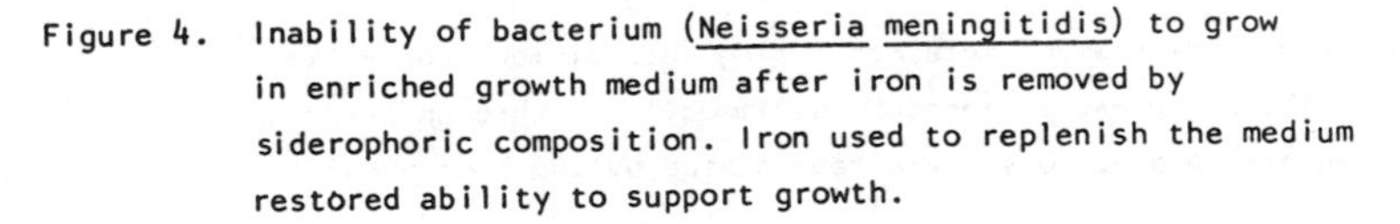
Figure 4. Inability of bacterium (Neisseria meningitidis) to grow in enriched growth medium after iron is removed by siderophoric composition. Iron used to replenish the medium restored ability to support growth.

Table 1

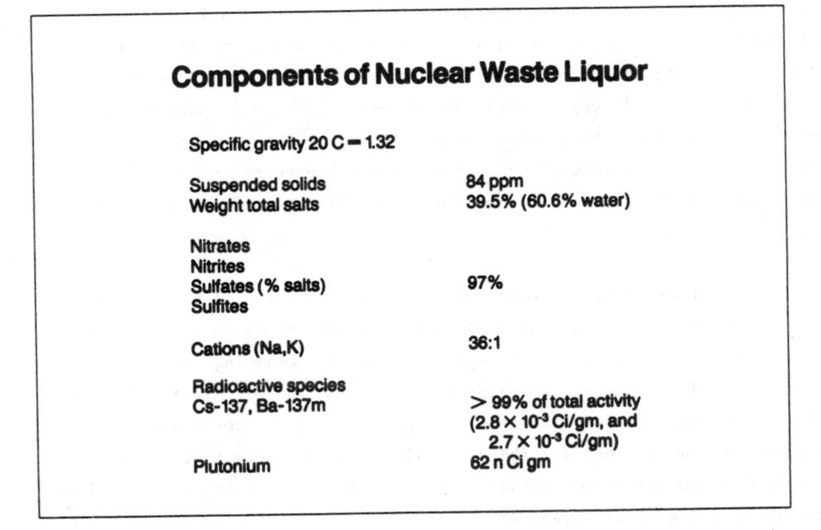

Components of Nuclear Waste Liquor

Specific gravity 20 C – 1.32

Suspended solids	84 ppm
Weight total salts	39.5% (60.6% water)
Nitrates	
Nitrites	
Sulfates (% salts)	97%
Sulfites	
Cations (Na,K)	36:1
Radioactive species	
Cs-137, Ba-137m	> 99% of total activity (2.8×10^{-3} Ci/gm, and 2.7×10^{-3} Ci/gm)
Plutonium	62 n Ci gm

Table 2

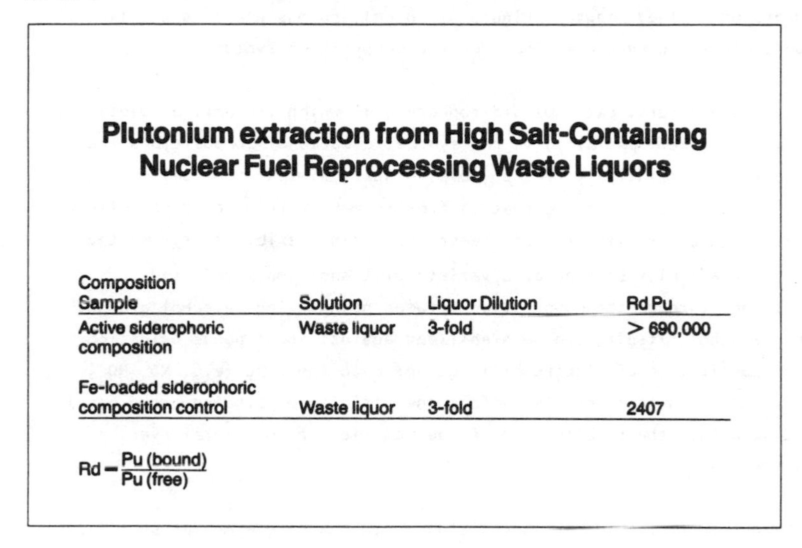

Plutonium extraction from High Salt-Containing Nuclear Fuel Reprocessing Waste Liquors

Composition Sample	Solution	Liquor Dilution	Rd Pu
Active siderophoric composition	Waste liquor	3-fold	> 690,000
Fe-loaded siderophoric composition control	Waste liquor	3-fold	2407

$$Rd - \frac{Pu\ (bound)}{Pu\ (free)}$$

In these studies the compositions removed plutonium from this complex solution with an Rd of >690,000 (undetectable). These results were encouraging in that other products tested reached Rd values in the range of only 30 to 300.

Iron in natural siderophores is held tightly through coordination bonds to either the N-substituted hydroxomate moieties of the Desferrioxamine or the catechol moieties of the Enterobactin (Figures 1 and 2). The active moieties in each instance are covalently bound to a structurally defined backbone of their respective ligands providing the rigid spatial configuration which is optimum for coordination bonding between the active moieties of the ligands and Fe III. This spatial orientation imposed on the active moieties by the chemically defined backbone to which they are attached aids in the selective "fit" of Fe III within these natural ligands. Although the natural siderophores have a predilection for iron, they are known to bind some other metals but with lower binding affinities.

DeVoe and Holbein (7) set out to synthesize a series of metal-capturing compositions with catechol, or substituted catechols, as the active moiety. Such compositions should have similar properties to those of Enterobactin.

Catechol was covalently bound to solid surfaces with bifunctional linking agents of defined lengths (Figure 3). Highly porous glass is a solid substrate found very practical for the composition synthesis.

Unlike the natural catachol siderophores, in which the active moieties are very much fixed spatially, in the synthetic catechol compositions the active moieties are relatively free to move in space similar to a balloon held fast by a length of string that is free to move within the zone defined by the length and flexibility of the string. This flexibility allows the "fit" and high-affinity binding of a variety of transition metals and actinides through the cooperation between neighboring catechol groups. The catechol, itself, can be stabilized against spontaneous oxidation by the substitution of electrophilic groups onto the ring (e.g. NO, NO_2, Cl, Br). Such substitutions also affect the affinity of binding among metals and so affect the predilection of the moieties for one metal over another (7).

Subsequent efforts (8) at composition syntheses were aimed at specific practical problems of industry. First efforts (9) were aimed at metals commonly present in nuclear waste, i.e. fission by-products such as Co-60, Sr-90, and Cs-137 (Table 3). The results of field trials demonstrated that indeed the new compositions could provide a practical means to remove radioactive metal from water.

Table 3

Evaluation of DeVoe-Holbein Compositions for the Removal of Radionuclides from mixed solutions[a][b]

Streams	Solution	Result (after treatment)
1. Primary Cooling Water	2000 ppm B, 5 ppm Li, 5 ppm Fe, 1 ppb Ni, 0.4 ppb Cr, 50 ppt Co-60 20 ppt Cs-137, 50 C	Co-60 undetectable Cs-137 undetectable (Rd) 4×10^6
2. Fuel Bay Water	Cs-137, Co-60, Sr-90 (Concentrations vary)	All undetectable
3. Citrate complexes	Fe-59, Co-60	Both undetectable
4. Floor drain Radwaste (b)	Co-60, Sr-90, Cs-137, Cr-51, Nb-95, Sb-125, Zn-65, Fe-59	All removed > 99.9%

(a) All compositions regenerable
(b) Co-60 0.1 ppt: Non radioactive Na 1000 ppm, Ca 173 ppm, Mg 90 ppm, B 22 ppm. Total Suspended solids 1000 ppm, pH 7.8.

A second area given special attention was the removal of non-radioactive toxic heavy metals from industrial waste streams (12,15). Results presented in Table 4 and 5 show the efficient removal of the target metals from a variety of actual industrial waste waters.

Table 4

The efficient capture of some toxic heavy metals using DeVoe-Holbein compositions

Source	Target Metal	Influent Concentration (ppm)	Effluent[a, b] Concentration (ppm)	Capture[c] Efficiency (%)
Plating effluent	Zn	1070	< 0.01	> 99.99
Plating effluent	Ni	670	< 0.01	> 99.99
Plating effluent	Cu	221	< 0.01	> 99.99
Plating effluent	Cr(III)	144	< 0.1	> 99.5
Plating effluent	Cr(VI)	3.8	< 0.4	> 99.5
Photographic effluent	Ag	10	< 0.04	> 99.5

(a) Effluent concentration at or below normal detection limits using either radioactive tracer or atomic absorption determinations
(b) Typically 100 cycles can be expected
(c) Capture efficiency determined as % reduction in influent concentration

Table 5

The efficient capture of some toxic heavy metals using DeVoe-Holbein components

Source	Target Metal	Influent Concentration (ppm)	Effluent[a, b] Concentration (ppm)	Capture[c] Efficiency (%)
Horticultural waste	Ag	20	< 0.06	> 99.5
Cooling tower bleed off/boiler blowdown water	Cr(VI)	694	< 0.4	> 99.9
High volume Incinerator scrub water containing trace heavy metals	Hg	0.9	< 0.001	
	Hg	100 (ppb)	< 0.001	
Contaminated groundwater	Hg	0.123	< 0.001	

(a) Effluent concentration at or below normal detection limits using either radioactive tracer or atomic absorption determination
(b) Typically 100 cycles can be expected
(c) Capture efficiency determined as % reduction in influent concentration

It was soon found that for industrial applications the DeVoe-Holbein compositions could benefit from specially designed water treatment plants in order to take full advantage of the special features of the compositions over those of conventional technology (Table 6). Ongoing field results from commercial installations were and continue to be excellent. Three examples (3,8) will demonstrate the practicality of this new process. Figure 5 shows an effluent stream from an industrial zinc plating line before and after treatment. Note the variable input, whereas the treated water has less than 0.01 ppm of the Zn.

Table 6

Special features of DeVoe-Holbein compositions

1. Selectivity	— Selective for target metals or groups of metals
2. Regenerability	— Regeneration cycle can be accomplished in 1.0-4.0 bed volumes with metal concentration reaching or exceeding 50,000 ppm off the column.
3. Conservation	— Conditioners and regenerants can be re-used requiring only occasional supplementation.
4. Physical stability	— No swelling or shrinking at extreme pH ranges or from heat.
5. Volume reduction	—Due to selectivity, volume reduction of metals can be orders of magnitude greater than ion exchangers.

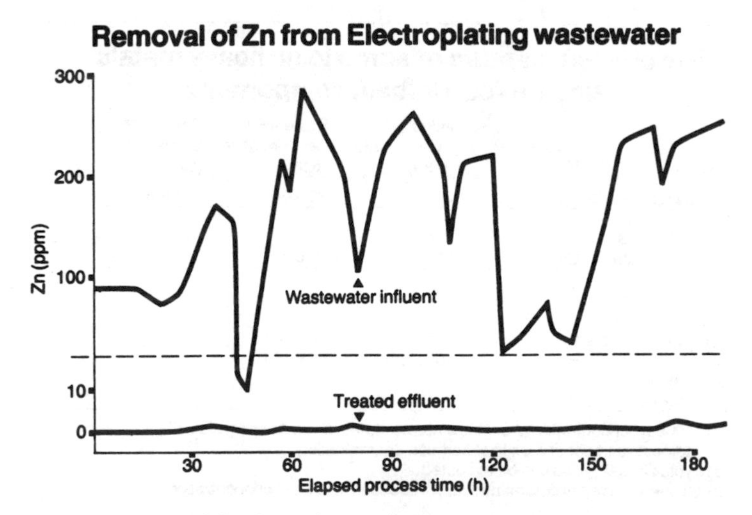

Figure 5. Removal of Zn from electroplating waste water.

The other metals in the plating rinse water (Table 7) did not interfere with the uptake of zinc, and the captured zinc was returned to the plating bath. Similar results (3) were obtained from a nickel plating rinse water (Table 8, Figure 6). Recycling of the composition beds,

Table 7

Zn-wastewater characteristics

element	Concentration (ppm)
Zn	192.6
Mn	0.05
Mg	8.54
Na	43.4
Sr	0.16
B	18.4
Si	1.2
Ca	39.9
pH	~ 6.3-7.3

Table 8

Ni-rinsewater characteristics

Element	Concentration (ppm)
Ni	136.6
Zn	0
Fe	1.75
Mn	0.26
Mg	3.06
Cu	0.08
Na	71.7
Sr	0.07
B	14.8
Si	2.9
Ca	14.1
pH 6.5	

usually with dilute mineral acids, was carried out in a short time period and the nickel at 8000 ppm was returned to the plating bath. In each example, treated water was re-used or discharged directly to the sewer.

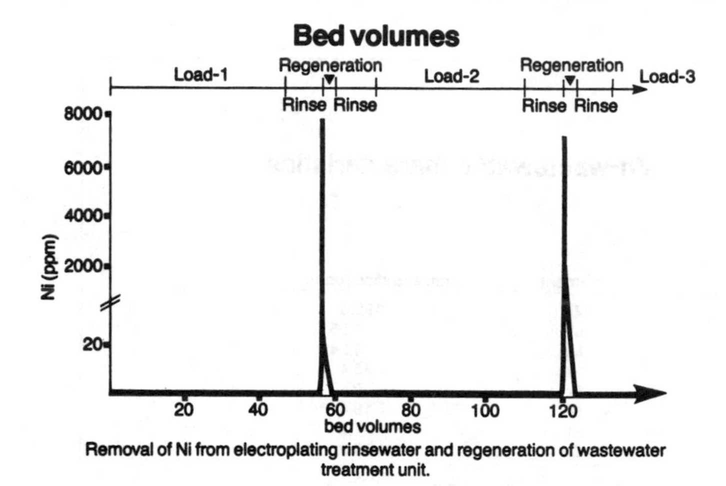

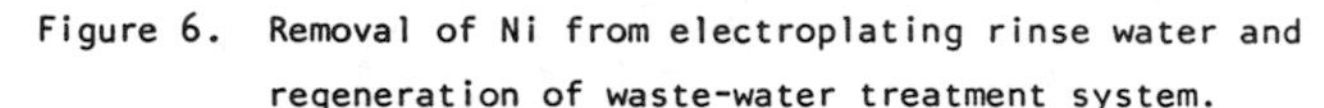
Figure 6. Removal of Ni from electroplating rinse water and regeneration of waste-water treatment system.

A third industrial example comes from installations in graphic or photographic processing centers. In this instance the rinse water contains 10 - 200 ppm of silver complexed to thiosulfate. The toxic nature of silver ion (roughly similar to cadmium) is often overlooked. In many countries silver in effluent streams is viewed only for its precious metal value. Results from several commercial units in the field show silver removed from photographic rinse water to concentrations less than 0.06 ppm, i.e. concentrations which exceed standards throughout the world. The composition bed captures approximately one ounce of silver per kilogram of composition. Recently, variations on the technology have been applied to the extraction and irreversible capture of gold from recalcitrant ore tailings, e.g. roasted arsenical pyrites.

In closing, a new technology, initially patterned after the highly efficient metal handling capabilities of biological systems, has progressed from the laboratory to industrial applications. This technology, the heart of which is a series of metal-capturing synthetic compositions, has application to nuclear waste, toxic waste and metal processing. Target metals can be removed from water to levels that meet world-wide environmental standards.

1. F.S. Archibald and I.W. DeVoe. 1979. FEMS Microbiol. Lett. 6:159.

2. F.S. Archibald and I.W. DeVoe. 1980. Infect. Immun. 27:322.

3. D. Brener and C. Greer. 1986. Novel synthetic compounds for the efficient removal of nickel and zinc from plating effluent. 7th American Electroplaters Society Florida/U.S. Environmental Protection Agency Conference on Pollution Control for the Metal Finishing Industry. January 27-29, 1986.

4. I.W. DeVoe 1982. Microbiol. Rev. 46:162.

5. I.W. DeVoe 1983. "Anregungen zur Forschung über die Pathogenität von Neisseria meningitidis - Überlegungen jenseits der Antibiotika-Therapy." in Infektiologisches Kolloguium I, Neues von "alten" Erregern und neue Erreger, Ed. C. Krassemann, Walter de Gruyter, Berlin 1983. pg 11-18.

6. I.W. DeVoe and B.E. Holbein. 1983. 4th Annual Conference Canadian Nuclear society, Montreal, Canada pg 1-26.

7. I.W. DeVoe and B.E. Holbein. 1985. Insoluble Chelating Compositions. United States Patent No. 4,530,963 July 23, 1985.

8. I.W. DeVoe, E. van der Vlist, and B.E. Holbein 1985. DeVoe-Holbein Technology: new technology for the extraction of hazardous waste metals. Hazardous Materials Management Conference, Hamburg, June.

9. C.W. Greer, D. Brener, E.N.C. Browne, I.W. DeVoe, B.E. Holbein, Roger Ek. 1985. "Vitrokele™ compositions: Novel, high affinity, metal selective and regenerable media for the removal of radioactive metals from aqueous nuclear waste streams". Waste Management '85, Tucson Arizona, 24-25 March 1985.

10. B.E. Holbein 1980. Infect. Immun. 29:886.

11. B.E. Holbein 1981. Infect. Immun. 34:120.

12. B.E. Holbein, I.W. DeVoe, L.G. Neirinck, M.F. Nathan, R.N. Arzonetti. 1984. DeVoe-Holbein technology: new technology for closed-loop source reduction of toxic heavy metals wastes in the nuclear and metal finishing industry. Ed. R. Clark. Massachusetts Hazardous Waste Source Reduction Conference Proceedings, Bureau of Solid Waste Disposal, October 17, 1984, pg 66.

13. L. Mason, B.E. Holbein, and F.E. Ashton. 1982. FEMS Microbiol. Lett. 13:187.

14. J.B. Neilands 1981. Ann. Rev. Biochem. 50: 715.

15. L.G. Neirinck and B.E. Holbein. 1985. Removal of heavy metals from waste streams with novel, high-affinity, selective and regenerable media. Annual Meeting of American Electroplaters Society, Detroit, Michigan, June 1985.

16. C. Simonson, D. Brener, and I.W. DeVoe. 1982. Infect. Immun. 36:107.

The Application of Carbon Adsorption Technology to Small-scale Operations for the Recovery of Gold from the Tailings of Old Mine Workings

By G. Woodhouse

BABCOCK WOODALL-DUCKHAM LIMITED, THE BOULEVARD, CRAWLEY, SUSSEX RHIO IUX, UK

Introduction

There are, scattered around the world, a large number of old abandoned gold mines. Often, the processes by which the gold was extracted from the ore were primitive by today's standards. Consequently, there are distinct possibilities that payable gold can be extracted from the tailings dams left by those earlier workings. Because the quantities of gold contained in such tailings are generally small, and also because they are often located in isolated areas to which access is difficult, it is essential that the process used to extract the residual gold should be simple yet efficient.

Conventional cyanide based processes are relatively expensive, both to instal and operate, because of the high cost of filtering the leach residue from the leachate. The advent of carbon adsorption based processes, such as carbon-in-pulp (CIP), carbon-in-leach (CIL) and adsorption from percolation leach solutions, which are much cheaper to instal and operate than conventional cyanide based processes, should enable many of these old tailings dams to be reworked economically. The income generated can either be an end in itself, or, if conditions are favourable, can be re-invested by opening up the old mine workings to provide more ore to feed the plant.

Background to the Gold Extraction Processes Available

Conventional cyanide based processes for gold extraction consist of the following essential steps[1]

RUN-OF-MINE ORE

|

CRUSHING & GRINDING

|

CYANIDE LEACHING
(CYANIDATION)

|

FILTRATION

|

ZINC PRECIPITATION

|

CALCINATION & SMELTING

|

BULLION

The purpose of crushing and grinding is to reduce the ore particle size sufficiently so that the gold is accessible to leaching. Typically the ore is ground to 70% less than 200 mesh (Tyler). The grinding process universally adopted is wet-milling, from which a slurry or "pulp" of finely ground ore in water is produced. Typical pulp density fed forward to cyanidation is 45% w/w solids.

In cyanidation, the gold content is leached out by an alkaline cyanide solution which is formed by adding lime and sodium or calcium cyanide to the pulp. Because the gold content of run-of-mine ore is generally low (normally less than 10 g/tonne) only relatively small amounts of reagents are required (typically 1.0 kg/tonne CaO and 0.15 kg/tonne NaCN). The leaching process is normally carried out continuously by passing the leach pulp through a series of agitated vessels. Since oxygen is essential to the dissolution process, air is bubbled into the pulp.

The net result of cyanidation is the solubilization of the gold to form aurocyanide ions, the overall reactions being[2]:-

$$2Au + 4CN^- + O_2 + 2H_2O \longrightarrow 2Au(CN)_2^- + H_2O_2 + 2OH^-$$

and/or

$$4Au + 8CN^- + O_2 + 2H_2O \longrightarrow 4Au(CN)_2^- + 4OH^-$$

Certain so-called "refractory" ores are not amenable to direct cyanidation because the gold particles are very fine and are occluded by various minerals - generally sulphides - within the ore. Such ores are termed refractory because the gold can often be rendered accessible to leaching by pre-roasting. It is not normally economic to roast run-of-mine ore, but since the gold is associated with certain minerals such as pyrites or arseno-pyrite, the ore can be concentrated by flotation, and roasting restricted to the concentrate. Alternatively, if the ore mineralogy is such that a satisfactory flotation concentrate cannot be produced, then pressure oxidation of the milled pulp can be used to oxidise the sulphides and release the gold for subsequent cyanidation[3].

After cyanidation, the leach solution is removed from the ore residue by filtration, generally by a number of very large rotary drum vacuum filters or alternatively by belt filters. The residue cake is washed to recover as much leach solution as possible before being disposed of.

The filtrate is then clarified, generally by pre-coated filters, and then vacuum de-aerated - these operations being an essential pre-requisite of zinc precipitation. Zinc dust and sufficient lead nitrate to form a zinc-lead couple are added to precipitate the gold. After removal by filtration, the gold slime is sulphuric acid washed to remove excess zinc, re-filtered, calcined and finally smelted with various fluxes such as borax and silica to produce bullion for subsequent refining.

In the CIP process, filtration of the leach solution from the ore residue is deleted. Instead, the aurocyanide content of the leached pulp is adsorbed onto activated carbon granules. These can subsequently be separated from the residue pulp by screening, because the carbon granules at 6-16 mesh (Tyler) are much larger than milled ore particles. In this way, the gold content can be transferred to the carbon at a concentration which is typically a thousand-fold higher than in the original ore.

In the CIL process, CIP is taken a stage further by combining the separate operations of cyanidation and adsorption into one.

Alternatively, instead of leaching the ore as a pulp, the leach liquor can be percolated through the ore (heap or vat leaching) and the gold adsorbed onto carbon from the leachate.

Various methods are available for the subsequent recovery of the gold from the carbon. These range from simple calcination in air, which of course results in the loss of the carbon, to several different methods of eluting (stripping) the gold from the carbon, so that it can be re-used. The gold can then be recovered from the eluate by electrowinning or, unusually, by zinc precipitation.

Although the ability of carbon to adsorb gold from cyanide solutions was discovered in 1894[4], it was not until relatively recently that the first large scale CIP plant was installed at Homestake's Lead Mine in South Dakota[5]. Since then, carbon adsorption based processes have been rapidly accepted as the modern methods for recovering gold and a considerable number of plants have been constructed throughout the world. However, perhaps because the technology is relatively young, and also because of varying mineralogy and other circumstances, the installed plants display a wide range of methods of utilising carbon adsorption. The more important of these process options are summarised in Figure 1.

It is the purpose of this paper to examine the various process options available particularly as regards their impact on small operations for the re-treatment of old tailings dams. Because of the large differences both in the scale of operation and in the gold content of the material to be processed, it is not normally economic to simply scale down a fully integrated plant designed to process a high tonnage of mined ore to the smaller scale required of most tailings re-treatment facilities.

Fig 1
PROCESS OPTIONS FOR GOLD RECOVERY USING ACTIVATED CARBON

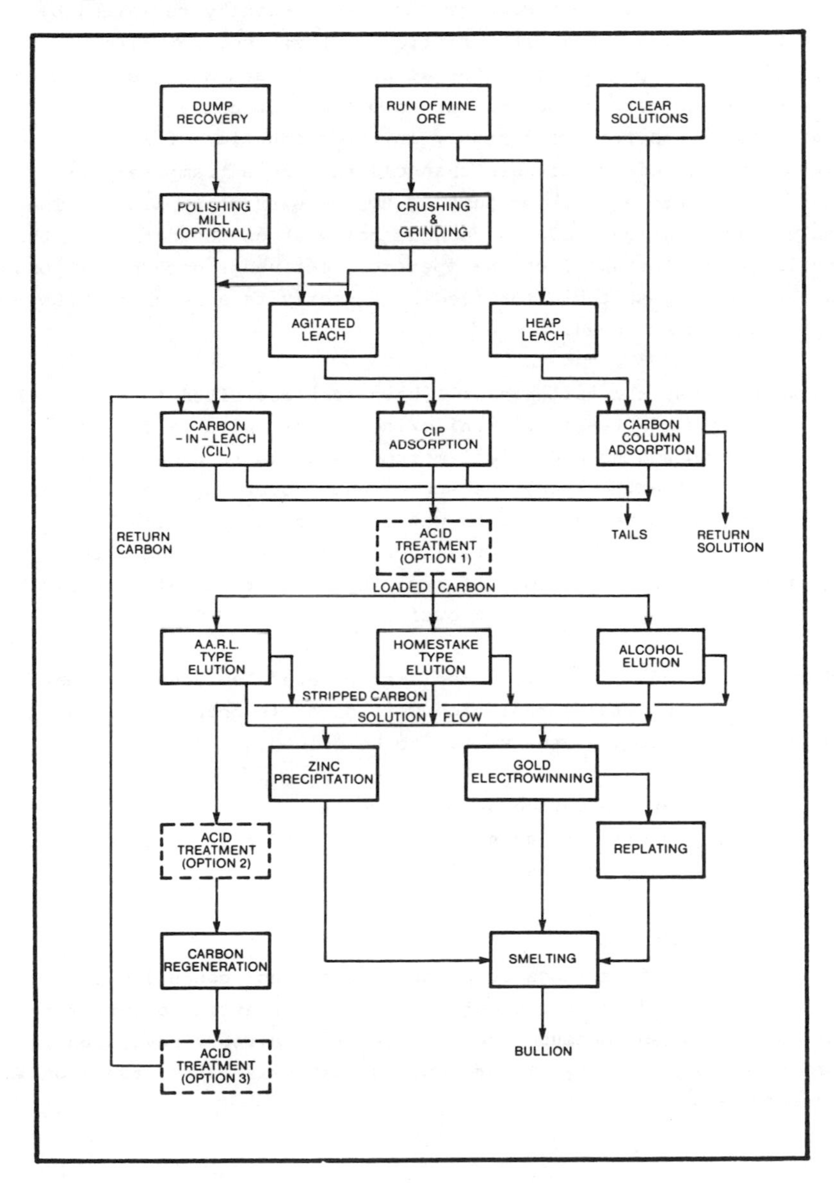

Reclamation of the Tailings

In the majority of cases, the old tailings dam will have been built up to form an artificial hill which is above the level of its surroundings. Such dams are most conveniently reclaimed by water monitoring. A monitor similar to that used for fire-fighting is supplied with water at about 14 bar gauge pressure and the resulting jet is used to cut into the soft compacted tailings. This results directly in the production of a pulp, which can be led by suitable channels through a tramp removal screen and into a sump for pumping to the extraction plant. The pulp concentration produced depends somewhat on the nature of the tailings and the skill of the operator, but under good conditions a 45% w/w pulp suitable for feeding directly to a leach or CIL plant can be achieved.

In some cases, the tailings will have been deposited in a natural or artificial depression. Monitoring is not suitable for such applications and so mechanical methods such as front-end loaders are used instead. Often, the subsequent processing plant requires that the solids be pulped with water. This is conveniently achieved by discharging the tailings into a tank fitted with a high capacity recirculatory slurry pump. The tank inlet should be fitted with a coarse tramp removal screen.

Mechanical means of tailings reclamation are also used at those facilities employing percolation methods of leaching, since such plants do not require the solids to be pulped.

Irrespective of the method adopted, reclamation is always a day-time only operation and as such presents buffer storage problems to the subsequent processing plant.

Solids Pre-treatment

The economics of tailings re-processing are not conducive to expensive methods of pre-treatment, such as roasting or pressure oxidation. Also, because the material will already have been ground during its original processing, extensive size reduction is unnecessary.

However, a light grinding of the tailings in a ball-mill can result in the achievement of significant improvements in subsequent extraction. These improvements are probably due as much to the removal of adhering clays and other fine debris as to actual size reduction of the particles. At one such installation, the monitored tailings were first pumped to a holding vessel, which was used to continuously feed a milling circuit to which lime and cyanide were added. This "milling in cyanide" process effected dissolution of a considerable proportion of the gold values. The milling circuit discharged to a thickener to increase the rather low solids content achieved by its particular monitoring operation, and the resulting excess cyanide solution was passed through a multi-bed carbon column in which its gold content was adsorbed. The thickened, partly leached pulp was then fed forward to a leach and CIP plant for recovery of its remaining gold values.

A pre-treatment step which is essential to subsequent CIP or CIL operations is pre-screening of the pulp. This is because the carbon is retained within the CIP or CIL vessels by static screens which allow the pulp to pass through. Any material of a similar size to, or larger than, the carbon will foul these static screens, and will also contaminate the loaded carbon. For this reason, the pulp should always be pre-screened at one standard mesh size smaller than that used subsequently for the carbon. Vibrating screens using slotted polyurethane decks have proved the most satisfactory for this service. Pre-screening should be carried out as early as practicable because much of the over-size material present in old tailings dams is of vegetable origin. Such organic materials readily adsorb gold from leach solutions, and consequently if pre-screening is not carried out until after leaching, then gold can be rejected with the over-size.

If percolation leaching is to be used, but the material is too fine, then agglomeration may be beneficial. The presence of a large proportion of sub 100 mesh (Tyler) material will inhibit efficient percolation[6] and cause channelling and other problems. Various methods of agglomerating fine gold bearing materials so that they are more amenable to percolation leaching have been tried. Water plus lime and/or cement are normally used as binders. Investigations by the United States Bureau of Mines showed that optimum agglomeration of a clayey gold ore was achieved at 12% moisture by adding 5 kg/tonne (ore) of Portland cement and then curing for 8 hours[6]. A significant improvement in the subsequent leaching time was achieved by agglomerating with an 8.6% w/w NaCN solution instead of water. Drum, pan or belt agglomerators[7] can be used and, on a small scale, converted cement mixers[8] have produced useful results.

Leaching

Conventional "pulp" leaching can be conducted either batchwise or continuously. Although continuous leaching is nearly always used for plants processing mined ore, for tailings re-treatment batch leaching offers certain advantages. In particular if, as is likely, the material can be effectively leached in, say, 16-20 hours, then it is convenient to load batch leach vessels from day-time reclamation operations, and allow leaching to proceed through the night ready for discharge next day. Also the reagents, lime and cyanide, can be added as weighed batches from sacks or drums. The only disadvantage of batch leaching is that, if it is followed by a continuous or semi-continuous CIP operation, then buffer storage of the leached pulp is required. However, continuous leaching requires a similar size of buffer storage, but before rather than after leaching. It also requires continual round-the-clock monitoring and continuous reagent addition systems.

Whether batch or continuous leaching is employed, compressed air agitation in conical bottomed leach vessels is preferable to mechanical agitation. This is particularly so in remote areas because of the maintenance and spares requirements of mechanical agitators.

To improve the degree of gold extraction obtained from certain ores, JCI[9] have developed a pipe reactor in which cyanide leaching is conducted under pressure - typically 48 bars. Their patent describes testwork on a stibnite rich material from which an improvement in gold extraction from 42.6% in 24 hours at atmospheric pressure to 90% after 2 hours in the pipe reactor. Such an operation would require high pressure slurry pumps and air compressors and, being a continuous process, buffer storage of reclaimed material. Some of these disadvantages could perhaps be circumnavigated by using batch leaching at elevated pressure instead of the pipe reactor.

An alternative to "pulp" leaching is some form of percolation leach. This can be either heap or vat leaching. Heap leaching requires the careful construction of a heap of material on a prepared drainage pad. Heaps of 1000 to 10 000 tonnes are commonly used[10]. Vat leaching is conducted in specially constructed concrete ponds[8]. In both cases the material is loaded dry and so percolation leaching is not suitable for materials requiring pre-milling. The heap or vat contents is then sprayed intermittently with alkaline cyanide solution and eventually water washed to displace the leachate. The overall cycle time is typically 15 to 17 days for vats[8] and 7 to 30 days for heaps[10]. Percolation leaching was initially developed and is mostly used for coarsely ground free-leaching ores. Heaps constructed of fine materials are prone to poor extraction due to channelling and in extreme cases to blinding which results in the leachate being discharged from the sides instead of from the base. Vat leaching of fine materials suffers from similar problems. Agglomeration of fine materials can be used to improve operations but adds significantly to the cost.

CIP Adsorption

Figure 2 illustrates a typical continuous leach and CIP adsorption circuit. The leached pulp is fed to a cascade of agitated vessels through which the leached pulp flows by gravity. In each vessel, the pulp is contacted with carbon granules which absorb the gold from solution. The pulp overflows from one vessel to the next via air-cleaned static screens, through which the carbon granules cannot pass. Pulp residence time per vessel is about 30-60 minutes[11]. Periodically, a proportion of the carbon inventory of each vessel is transferred to its neighbour up the cascade by air-lifting a quantity of carbon-bearing pulp. Highly loaded carbon is removed from the uppermost vessel by air-lifting pulp from it onto a vibrating screen. Fresh or stripped carbon is added to the lowest vessel to maintain the carbon inventory at typically 25 g/litre[11] of residence volume. Because adsorption operates counter-currently, it is possible to reduce the soluble gold content of the pulp to very low levels (typically 0.1% of original) whilst at the same time producing highly loaded carbon.

The optimum number of vessels in the cascade depends on many factors such as the soluble gold content and adsorption kinetics - which vary from ore to ore because of the adverse effects of organic and certain metallic contaminants. For a re-treatment operation, 4 or 5 vessels would normally be sufficient.

In general, there are few differences between a large-scale CIP adsorption circuit operating on mined ore and a small unit processing tailings. One exception is vessel agitation which should preferably be by air rather than by the mechanical means shown, even though some drop in adsorption efficiency[11] may result. Also carbon feeding and the removal of loaded carbon can be manual rather than mechanical operations. In order to achieve the advantages of counter-current adsorption, CIP must be operated continuously. A batch sequence type of operation could be devised but would be too complicated to operate efficiently. Continuous operation of course has the disadvantage of round-the-clock manning, but if CIP adsorption is employed then there is little alternative. One possible solution is to operate only during the day and to leave the vessels agitating during the night. Such a mode of operation was practised at one re-treatment facility, where the adsorption circuit operated in conjunction with batch leaching. However, the adsorption circuit must then be larger to accommodate the higher instantaneous throughput.

Adsorption from Solutions

Percolation leaching produces a virtually clear solution which contains the extracted gold values. Carbon adsorption is the most common approach for recovering these values. In virtually all such installations, the solution is passed up through carbon beds at such a velocity that the carbon is slightly fluidised. As in CIP adsorption, some means of counter-current operation is required to load the carbon to a high gold level whilst reducing the soluble gold losses to as low a value as possible. This can be achieved by a multi-stage adsorption tower similar to that used for adsorbing uranium onto resin. In this device, fresh carbon is added to the top stage and by occasional short period flow reversals the carbon is moved down from stage to stage. Alternatively, separate open-topped tanks can be used, arranged in cascade so that the solution flows from one to the next by gravity. These suffer the disadvantage that the carbon must be

Fig 2
FLOW SCHEMATIC OF LEACH AND CIP ADSORPTION CIRCUITS

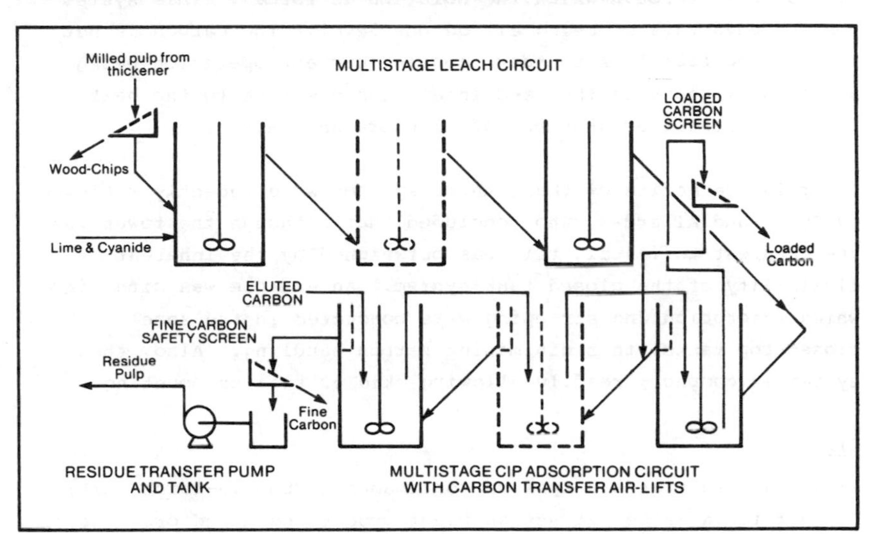

Fig 3
FLOW SCHEMATIC OF CIL LEACH-ADSORPTION CIRCUIT

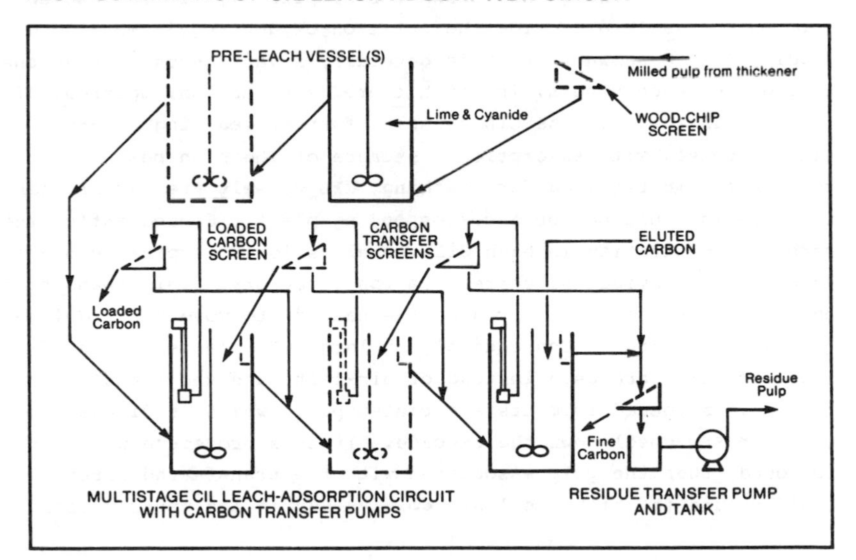

moved uphill using eductors. A third option is to use separate closed tanks through which the solution is forced. This system has the advantage of being all on one level. The carbon is not transferred from tank to tank, counter-current operation being achieved by changing the feed inlet from one tank to the next using a suitable arrangement of pipework and valves.

The relative merits of these three systems were recently reviewed by Pyper and Allard[12], who concluded that although the tower was the cheapest to instal, this was outweighed by the inherent flexibility of the closed tank system. An example was cited in which adsorption and stripping were conducted in the same closed-top tanks, thus minimising carbon handling. Also, the system is compact, readily allowing skid or trailer mounting.

CIL

Figure 3 illustrates a typical continuous carbon-in-leach (CIL) circuit for a large extraction plant processing mined ore. As in conventional leaching, the pulp overflows from one vessel to the next, the leach reagents (lime and cyanide) being added to the first vessel. Although not shown for simplicity, air is bubbled into each vessel to provide the reaction oxygen requirements. Studies[13] have shown that it is generally advantageous to have one or two pre-leach vesels, in which there is no carbon, upstream of the CIL cascade. In the CIL vessels, further leaching occurs simultaneously with adsorption. Because of the high pulp residence time required for leaching, CIL vessels are larger than CIP vessels would be for a corresponding plant. Consequently, the carbon concentration in each CIL vessel is lower (typically 5 g/l) in order to provide the correct carbon inventory. This means that more pulp must be transferred up the cascade to move the required quantity of carbon. Consequently on large CIL plants, recessed impeller pumps are used instead of air-lifts and the transferred carbon is screened from its associated pulp, which is discharged to the next vessel down the cascade. If this procedure were not followed, then the pulp associated with the transferred carbon would dilute the higher soluble gold values in the upper vessel.

For a small dump re-treatment operation, some of the expensive complications of a large CIL plant would need to be dispensed with. For instance, mechanical agitation would be replaced by air and carbon transfer would be by air-lift rather than by pump. Also the extra complication of carbon transfer screens could be avoided by adding one or two additional CIL vessels to make up for the inefficiency caused by the "back dilution".

A possible simple alternative to continuous CIL is batch operation, which has been used on at least one re-treatment facility in South Africa. The pulp was loaded into an air-agitated vessel and the required amounts of lime, cyanide and carbon added. Leaching and adsorption occurred simultaneously during the night and next day the vessel was emptied through a screen to catch the carbon. In order to build up sufficient gold on it, the same carbon was used for several batches of pulp. Because there was no counter-current operation, the carbon could not be loaded to very high gold levels, and had to be replaced when about 1000 g/tonne Au was attained, in order to avoid excessive gold losses.

Separation of the Gold from the Carbon

As mentioned previously, the gold can be recovered from the carbon by burning it. Although this results in the loss of the carbon, for very small installations it may be the most economic solution. If the carbon is to be burnt, then it should be loaded to as high a gold content as possible, so as to maximise the added value. This consideration rules out such operations as batch CIL, which produce lightly loaded carbon.

In most cases the gold is eluted (or stripped) from the carbon, which can then be re-used. Because elution plants are quite complicated, it may be advantageous to have the carbon eluted at a nearby larger carbon based gold extraction plant, if one is available. Alternatively, several small tailings re-treatment operations could club together and build a joint elution facility. If a re-treatment plant has to have its own elution plant, then it should be designed as simply as possible. Consequently, such complications as pressure elution at above the normal boiling point of the eluate should be rejected in favour of operation at atmospheric pressure, even though the latter will result in a longer elution period[10]. The use of alcohol elution, as has been practised at a few installations[14,15], is generally undesirable because of the fire hazard.

Fig 4.
FLOW SCHEMATIC OF ELUTION AND REGENERATION CIRCUITS.

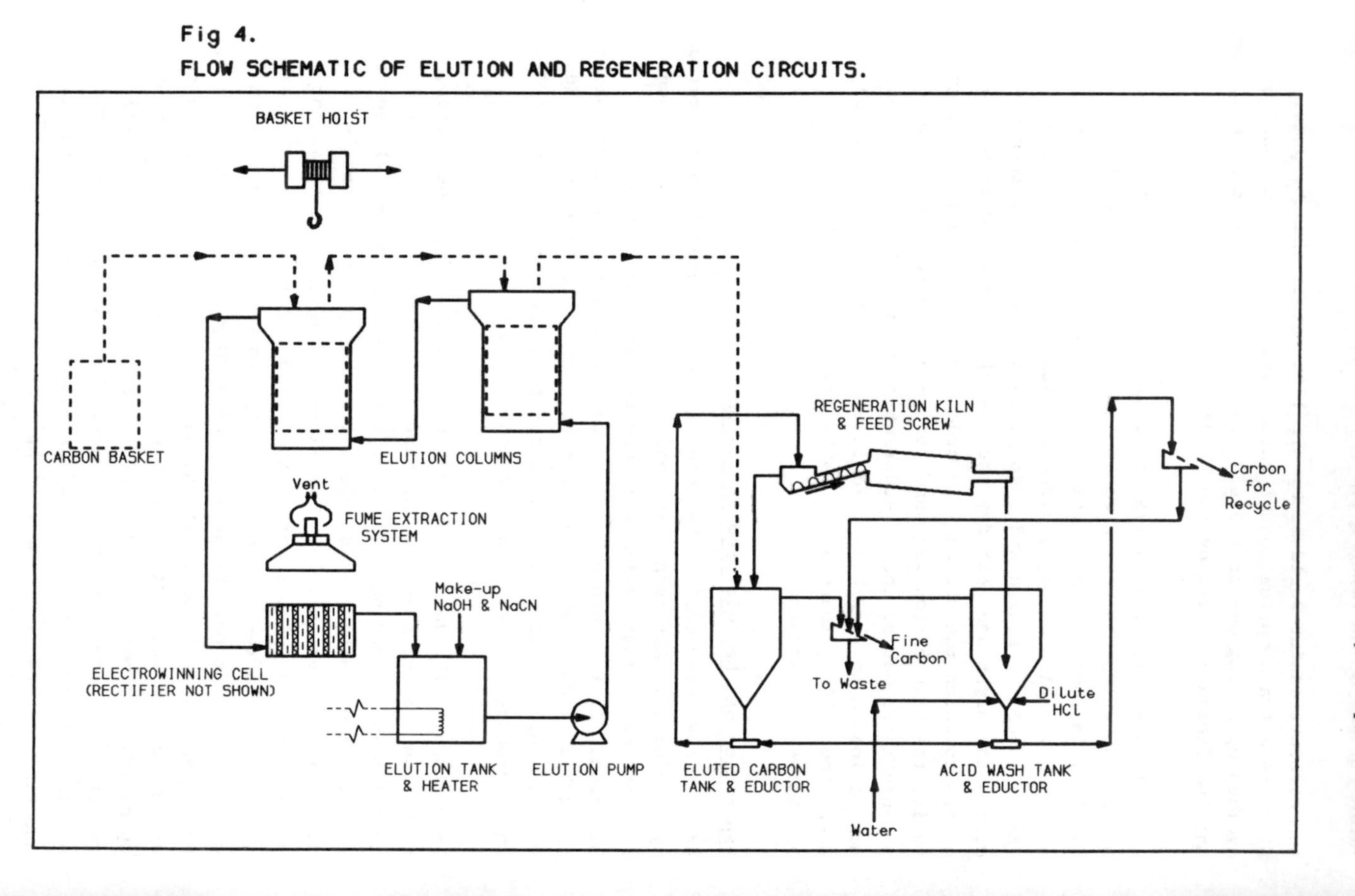

Figure 4 illustrates a typical elution and regeneration plant for a medium sized tailings re-treatment operation. The loaded carbon is placed in a perforated basket, which is lifted into the 1st elution column. The stripping solution is caustic sodium cyanide (1% NaOH, 0.2% NaCN) which is maintained at about 90°C by an electrical heater in the elution tank. The hot solution is circulated up through the elution columns so that the carbon granules are slightly expanded. The gold is stripped from the carbon and passes into the solution which then flows to the electrowinning cell. Various designs of cell have been developed. In most, the gold is deposited onto steel-wool cathodes. The type illustrated in Figure 4 was developed by MINTEK[16] in South Africa and consists of alternate perforated anodes and porous steel-wool cathodes, the solution flowing through the anode-cathode pairs from one end of the cell to the other, finally overflowing back to the elution tank. After 24 hours or so, the carbon baskets are removed. That from the 1st column is placed in the 2nd and that from the 2nd is emptied into the eluted carbon tank. The empty basket is then refilled with loaded carbon and placed in the 1st column. The same elution solution is recirculated continually, make-up reagents being added to maintain their concentrations. Impurities such as copper gradually accumulate in the solution and as a result it must eventually be replaced.

The eluted carbon is regenerated in an electrically heated kiln at about 700°C in an oxygen free atmosphere. Transfer to the kiln is by water using an eductor, the kiln feeder being a dewatering screw. Water charged with the carbon produces steam which displaces air from the kiln. The regenerated carbon is washed in dilute hydrochloric acid to remove calcium scale which would otherwise occlude the gold adsorption sites. After final washing and screening to remove fines, the carbon is ready for re-use.

For a very small operation, the complication of two elution columns is not justified, although a single column will require a longer elution time. Also, if possible, regeneration should be avoided. At some installations, it has been found that the benefits of regeneration are marginal, whereas at others it is essential. Pilot testwork can be used to determine whether regeneration is required or whether it would be more economic to occasionally purge the carbon. Such testwork must be realistic and include a complete adsorption-elution circuit using actual ore and water from the proposed site. The presence of organics in the water has been shown to be highly detrimental to adsorption and to greater increase the necessity of regeneration.

Gold Recovery Operations

Tailings re-treatment operations do not normally produce sufficient quantities of gold to warrant smelting and refining. Instead, the gold is sold at as high a concentration as practicable to a refinery.

Gold collected in electrowinning cells is recovered by removing the cathodes and dissolving their steel-wool content in hydrochloric acid. The insoluble gold-bearing residue is filtered off, dried or calcined and sold as a concentrate. Similarly, gold-bearing sludges which accumulate in electrowinning cells are filtered off and dried or calcined for sale. Fine abraided carbon which collects in, for example, the safety screens of CIP and CIL plants is calcined to burn off the combustibles and the resulting ash sold as a gold-bearing concentrate. Any other carbon rejected from the system is similarly calcined.

If the tailings were originally from amalgamation operations, then there is a risk of mercury accumulation in these concentrates. This can present a health risk during drying or calcination operations[8].

Comparison of the Various Process Routes

Because of the large number of site and ore related factors involved, it is not possible to make specific recommendations regarding the optimum treatment method for a particular material, without carrying out test-work and preparing an economic evaluation. However some general pointers can be drawn to assist in the development of preliminary flowsheets and in drawing up test-work programmes.

(i) Size and Richness of Reserves. The maximum return on investment will be realised by tailoring the process plant to suit the size, and the gold content, of the reserve of tailings. That is to say that the plant's throughput, and its level of sophistication, should be chosen to achieve the maximum net income over the life of the reserve. In general, plants processing tailings are of a lower level of complexity than plants processing large reserves of mined ore, and moreover, the lower the size and richness of the reserves, the simpler the plant must be to be economic. Such simplicity will result in a lowering of extraction efficiency, but for small reserves this will be outweighed by the savings in capital and operating costs.

Thus for small projects, simple less efficient processes such as batch CIL or percolation leaching should be examined, whereas for larger projects, continuous pulp leaching and CIP adsorption (or possibly continuous CIL) are more likely to yield the maximum return.

(ii) Site Specific Considerations. Some important pointers can be drawn from a consideration of site conditions. For instance, the topography of the tailings dam has a direct bearing on the best method to be used for reclamation. The presence or otherwise of a clear load-bearing level site close to the tailings is of obvious importance. Climate too is important. For instance, heap leaching could prove difficult in areas of very high rainfall or of severe frosts. Conversely, if water is scarce then the lower water demands of heap leaching may prove to be beneficial. The remoteness of the site and the availability of skilled labour are also factors of importance in deciding the type of plant to be used. Pointers can also often be drawn from the previous history of the mine and in particular the origin of the tailings.

(iii) Metallurgical Properties of the Material. A simple laboratory programme of bottle leach and adsorption tests will greatly assist in the preparation of preliminary flowsheets for evaluation purposes. Such tests can determine whether the gold leaches easily, or conversely the material is refractory; whether carbon can be loaded to high levels, or conversely that its adsorption performance is poor, perhaps because of the presence of other elements or organic materials which compete for the active sites; whether a slight grinding or attrition scrubbing is beneficial; whether CIL has advantages over CIP, because, for instance, that there are competing adsorbants present; whether the size analysis of the material is suitable for percolation leaching.

In many cases, such testwork will show that good extraction can be attained with simple processes and thus enable preliminary evaluation flowsheets to be drawn up. In other cases, e.g. where the ore proves to be refractory, then more sophisticated and more expensive testwork will be required, before such a preliminary economic evaluation can be completed. Once the preliminary evaluation has shown that the project should be viable, then pilot scale testing and a final evaluation should be conducted.

Acknowledgment

The author wishes to thank the Directors of Babcock Woodall-Duckham Ltd - a Babcock International plc company - for permission to publish this paper.

References

[1]R.J. Adamson (Editor), "Gold Metallurgy in South Africa", Chamber of Mines of South Africa, Johannesburg, 1972.

[2]N.P. Finkelstein, "Gold Metallurgy in South Africa, Chamber of Mines of South Africa, Johannesburg, 1972, Chapter 10, p.304.

[3]R.S. Kunter, J.R. Turney and R.D. Lear, "First International Symposium on Precious Metals Recovery", 1984, Paper XIV.

[4]W.D. Johnson, U.S. Patent 522,260 (1894).

[5]K.B. Hall, World Mining, 1974, November, 44.

[6]G.E. McClelland and J.A. Eisele, "Improvements in Heap Leaching to Recover Silver and Gold from Low-Grade Resources", USBM Report RI 8612, 1982.

[7]D.A. Mulligan, "First International Symposium on Precious Metals Recovery", 1984, Paper XIX.

[8]C. Authur, D. Jordan and C. Poole, "First International Symposium on Precious Metals Recovery", 1984, Paper XXIV.

[9]C.W.A. Muir and L.P. Hendriks, S.A. Patent Application 83/6301 (1983).

[10] H.J. Heinen, D.G. Peterson and R.E. Lindstrom, "Processing Gold Ores Using Heap Leach-Carbon Adsorption Methods", USBM Report IC 8770, 1978, p.7.

[11] P.A. Laxen, C.A. Fleming, D.A. Holtum and R. Rubin, "Proceedings, Twelfth Congress of CMMI", South African Institute of Mining and Metallurgy, 1982, Vol. 2, p.551.

[12] R. Pyper and S.G. Allard, "First International Symposium on Precious Metals Recovery", 1984, Paper XXIII.

[13] G.M. Newrick, G. Woodhouse and D.M.G. Dods, World Mining, 1983, 36, No.6, 48.

[14] Anon, Mining Magazine, 1982, July, 20.

[15] L.R. Todd and A.R. Anderson, "First International Symposium on Precious Metals Recovery", 1984, Paper II.

[16] A.O. Filmer, "Carbon-in-Pulp Technology for the Extraction of Gold", Australasian Institute of Mining and Metallurgy, 1982, 49.

Ion Transfer by Solid-supported Liquid Membranes

By M.M. Kreevoy

DEPARTMENT OF CHEMISTRY, UNIVERSITY OF MINNESOTA, 139 SMITH HALL, 207 PLEASANT ST. S.E., MINNEAPOLIS, MINNESOTA 55455, USA

Solvent Extraction

Extraction of ions, accompanied by chemical reaction or exchange of ionic partners, is an old technique, widely practised in industry and in analytical chemistry, and sometimes called liquid ion exchange. This sort of solvent extraction will be discussed in considerable detail because its chemistry is identical with that of liquid membranes. The commonest examples are two-phase acid-base reactions, as shown in eqs. 1-4. (An overbar indicates a substance dissolved in a water-immiscible solvent; other participants are in aqueous solution.)

$$Cu^{2+} + 2\overline{RH} \rightleftharpoons 2H^{+} + \overline{CuR_2} \quad (1)$$

$$UO_2(SO_4)_2^{\;2-} + 2H^{+} + 2\overline{B} \rightleftharpoons \overline{(BH^{+})_2\cdot[UO_2(SO_4)_2^{\;2-}]} \quad (2)$$

$$Co^{2+} + 2\overline{PH} \rightleftharpoons 2H^{+} + \overline{CoP_2} \quad (3)$$

$$\overline{Q^{+}A^{-}} + H^{+} + NO_3^{\;-} \rightleftharpoons \overline{Q^{+}NO_3^{\;-}\cdot HA} \quad (4)$$

The reagents used couple the typically non-spontaneous transfer of the target ion with a strongly spontaneous ion transfer to make the whole process feasible. The reagents must have considerable specificity for the target ions under the prevalent conditions. The equilibrium must be displaced to the right under the conditions of loading and to the left under the conditions of stripping. Chemical equilibrium has to be established fairly rapidly, and intractable emulsions have to be avoided. A typical example of the phenolic oximes used for copper extraction, and the complex it forms with Cu^{2+}, is shown.[1] These extractants have a high selectivity for Cu(II) over Fe(III) and Al(III) which commonly coexist with it. It appears to be important for selectivity that all the potential hydrogen bonding valences are satisfied in the complex.

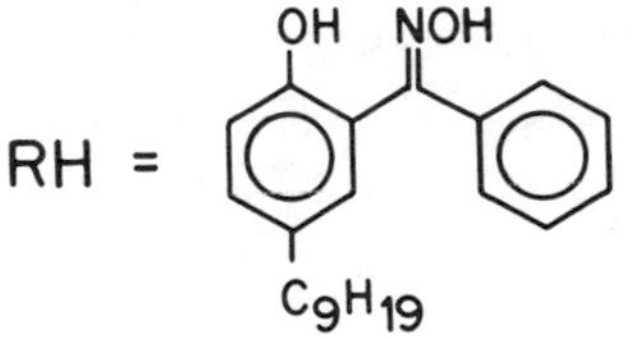

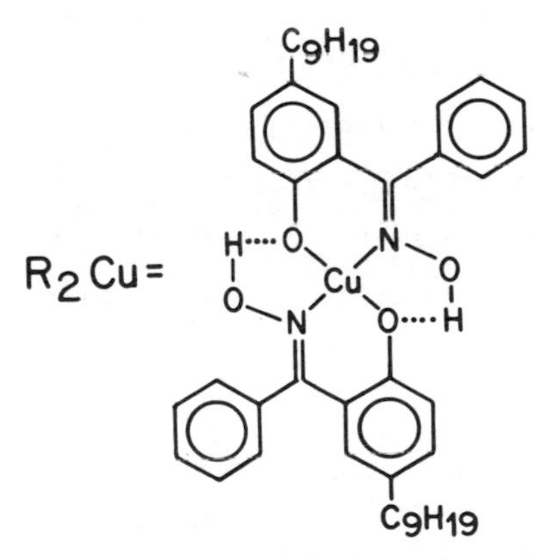

The most common neutral base used for uranium extraction has been a tertiary amine, Alamine[R] 336, the alkyl groups of which are half C_8H_{17} and half $C_{10}H_{21}$.[2] (Alamine is a registered trademark of Henkel America Inc.)

Dialkylphosphoric, phosphonic and phosphinic acids have been used as PH, for cobalt separation from nickel and other metals, with the phosphinic acids apparently exhibiting the greatest selectivity for cobalt.[3]

Nitrate, of course, is not a metal at all, but it is a common contaminant of domestic water, and its economic removal is a desirable objective.[4] Further, the same ideas which facilitate the decontamination of nitrate-polluted water are also likely to be helpful in the extraction of metals as anionic complexes. Nitrate can be transfered from an acidic feed to a basic strippant, using ordinary secondary or tertiary amines in hydrocarbon solvents as the extractants, in much the same way that uranium is isolated (eq.2).[5] However the equilibrium constant for extraction is modest under these conditions, and the feed must be quite acidic in order to extract practical amounts of NO_3^-.[5] This would make the water unsuitable for domestic use.

The position can be improved by using trioctylphosphate (TOP) as the solvent. It increases the equilibrium constant for extraction by something over two orders of magnitude.[6] TOP probably achieves this effect by stabilizing the ammonium cation through hydrogen bonding, as shown. This effect would also improve the extraction of uranium from very lean feeds, although there would be economic problems, as noted below.

The extraction of nitrate and a proton is an acid-base reaction, and its equilibrium constant can be increased by using a stronger lipophilic base. However quaternary ammonium hydroxides are unsuitable. They undergo elimination reactions at significant rates even at 25° C, and they also

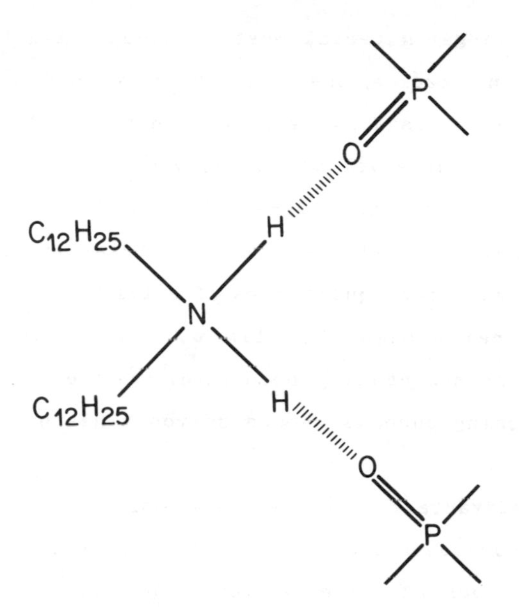

solubilize a good deal of water in the organic solvent. We have solved this problem by using quaternary ammonium salts with anions, A^-, that are more basic than neutral amines, but less basic than OH^-, and a good deal less hydrophilic. We have had good results with the phenate and sulfonamidate shown.[7] The phenol or sulfonamide which is formed on

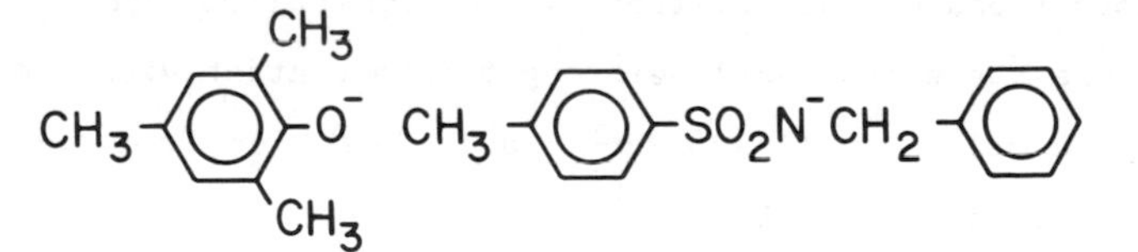

reaction of H^+ and NO_3^- with $\overline{Q^+A^-}$ is shown, in eq. 4, as hydrogen bonded to the ion pair. The extent to which this is actually the case is not known.

Because of the mechanical requirements of mixer-settlers, and because, in conventional solvent extraction, the target material must be accumulated in the organic phase during loading, the ratio of the aqueous to the organic volume cannot usually be more than about 10. This leads to the use of large volumes of solvent, particularly when the feed is lean, and severely limits the choice of solvents which may be considered for industrial practice. The fact that significant quantities of solvent may be lost by entrainment reinforces this limit. In industrial practice, kerosine has usually been used, and the potential advantage from using purpose-chosen solvents has not been realized.

Another disadvantage of conventional solvent extraction emerges from a consideration of its thermodynamics.[8] The condition of equilibrium for the extraction step is

$$\mu_{Cu^{2}(fd)} + 2\mu_{\overline{RH}(fd)} = \mu_{\overline{R_2Cu}(fd)} + 2\mu_{H^+(fd)} \tag{5}$$

or

$$a_{Cu^{2+}(fd)}a^2_{\overline{RH}(fd)} = a_{\overline{R_2Cu}(fd)}a^2_{H^+(fd)} \tag{6}$$

where μ is a chemical potential and a is an activity. Similar conditions define equilibrium for stripping. If the feed and the strippant could be brought into contact with the same organic solution H^+ and Cu^{2+} would transfer from the feed to the strippant until

$$a_{Cu^{2+}(fd)}a^2_{H^+(st)} = a_{Cu^{2+}(st)}a^2_{H^+(fd)} \ . \tag{7}$$

Thus Cu^{2+} could be driven from a 500 ppm feed to a 5% strip by a difference of one pH unit in their acidities.[8] Similar considerations apply to the other systems, outlined in eqs. 2-4. In fact, this ideal cannot be closely approached in conventional solvent extraction, because a considerable amount of Cu^{2+} must be stored in the form of R_2Cu between the extraction and stripping, so that the relation between the feed after extraction and the strippant after stripping is actually given by eq. 8.

$$a_{Cu^{2+}(fd)} a^2_{H^+(st)} a^2_{\overline{RH}(fd)} a_{\overline{R_2Cu}(st)} = a_{Cu^{2+}(st)} a^2_{H^+(fd)} a^2_{\overline{RH}(st)} a_{\overline{R_2Cu}(fd)} \qquad (8)$$

Since $a_{\overline{RH}(fd)} < a_{\overline{RH}(st)}$ and $a_{\overline{R_2Cu}(fd)} > a_{\overline{R_2Cu}(st)}$, eq. 8 demands a larger pH difference than eq. 7 or else requires a richer feed, or both. This is not presently a problem in the copper industry, but it will, inevitably, become so as leaner and leaner copper ores must be treated.

Supported Liquid Membranes

It is now technically possible to replace the mixer-settlers of conventional solvent extraction with a membrane, fabricated by filling the pores of a thin, porous, plastic support with the same sort of water-insoluble solutions described above.[9,10] The feed is circulated past one face of the membrane, where extraction occurs (the loading face); the target-carrier complex diffuses across the membrane; and stripping takes place at the other face (the stripping face), past which the strippant circulates. The nominal pore sizes of the membranes are, typically, 0.02-0.2 μm, but some of the supports in use — Goretex[R] for example — (Goretex is a registered trademark belonging to W.L. Gore Associates, Inc.) actually appear, under magnification, to be masses of fibers, rather than a continuous material with pores. Capillary forces hold the organic in the plastic support. Since the organic film is only 20-100 µm thick, the residence time of the ions of interest is usually only a few minutes, and the active reagent is promptly reused, so the reagent requirement is reduced by many orders of magnitude, compared to conventional extraction. Since almost any volume of feed can be contacted with almost any area of organic film, the requirement for organic solvent is small and does not rise strongly with the volume of aqueous feed to be treated. By using a large volume of feed and a small volume of an appropriate strippant, large increases in concentration can be achieved in a single stage. In addition, since the aqueous and organic phases are never intimately mixed, as they are in conventional extraction, loss of the organic phase by entrainment and the corresponding contamination of the aqueous streams can be minimized. The absence of a large inventory of the organic phase should minimize loss by spills, leaks, and other accidents. The reduction in required inventory makes a much wider range of reagents and solvents economically available for industrial processes. These advantages are obtained at a cost, however. The area of the membrane per unit volume of apparatus is limited, as opposed to the mixer-settlers, in which interfacial area is made very large by creating a dispersion. The limited area of the membrane demands very high specific rates of transfer if useful fluxes are to be obtained.

Also, presently available synthetic membranes, like conventional solvent extraction apparatus, are not able to operate effectively close to the condition of equilibrium, expressed in eq. 7, although living cell membranes can do so.

The reason is that present synthetic membranes are still thick by comparison with molecular dimensions. The target and its carrier must diffuse across the membrane, and a substantial concentration gradient must be maintained within the membrane to achieve a useful flux. This makes the conditions of loading substantially different from the conditions of stripping, as they are in conventional solvent extraction, and puts us back to eq. 8. The achievement of a synthetic membrane of molecular dimensions, which could maintain a useful flux under conditions within about an order of magnitude of equilibrium, remains a research goal.

Recent technological developments have brightened the prospects for membrane devices by increasing the available membrane area in a given physical volume.[11,12] Membrane area has been increased by mounting a large number of hollow fibers in a pipe, as shown in fig. 1. The feed flows down the interior of the fibers, which have common inlet and outlet chambers. In this way a packing factor of about 50 cm^2 membrane per cm^3 of device gross volume can be obtained.

Supported liquid membrane extraction of uranium, using the chemistry of the AMEX solvent extraction process,[2] summarized in eq. 2, has been successfully pilot-plant tested.[11] And, as noted, it is likely that further improvement could be achieved by specifically tuning the chemistry to the requirements of the membrane process. However the generally depressed state of the uranium market makes it unlikely that any new technology requiring substantial new investment can be put in service.

Solvent extraction is presently used in the refining of copper (eq. 1) thus it is also a prime target for supported

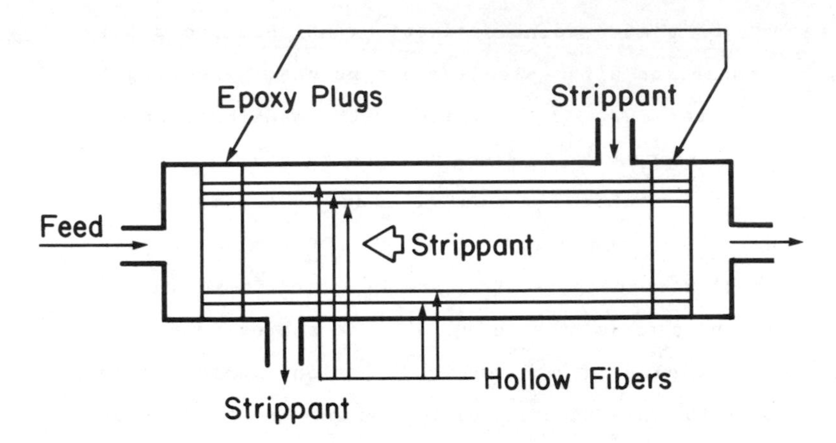

Figure 1. Pipe module hollow fiber membrane device.

---

liquid membrane processing. Here, however, an interesting scientific problem has arisen, in addition to economic problems similar to those of the uranium field. In operation a supported liquid membrane based on the chemistry of eq. 1 takes up water, and eventually becomes inactivated. This is shown by fig. 2. The uptake of water appears to be related to the transfer of copper. It does not occur if the same system is operated in the same way, but without copper. It seems likely that the transfer of water to the organic phase is an integral part of the mechanism by which Cu^{2+} is transferred to the organic phase.

The extraction of nitrate from water intended for domestic use or animal watering may beat mineral processing as the first large scale use of supported liquid membranes. In many ways this is an ideal application for supported liquid membrane technology. A small amount of nitrate is to

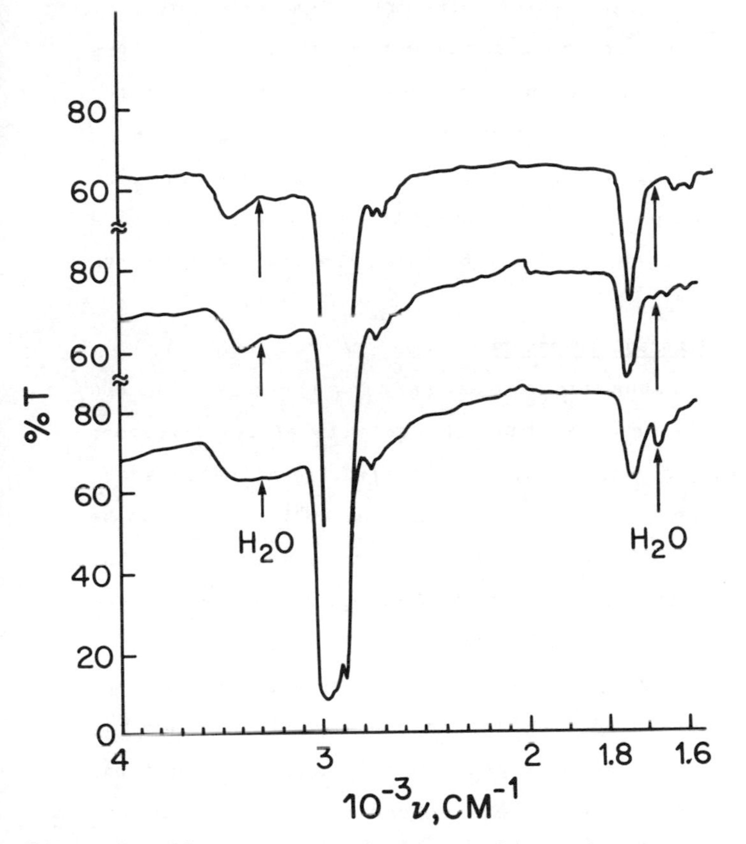

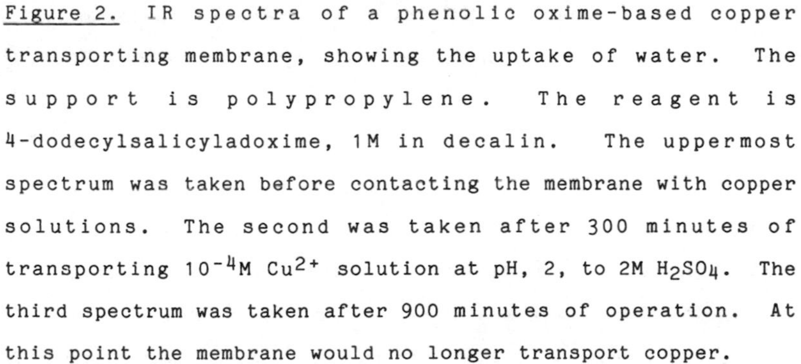
Figure 2. IR spectra of a phenolic oxime-based copper transporting membrane, showing the uptake of water. The support is polypropylene. The reagent is 4-dodecylsalicyladoxime, 1M in decalin. The uppermost spectrum was taken before contacting the membrane with copper solutions. The second was taken after 300 minutes of transporting 10^{-4}M Cu^{2+} solution at pH, 2, to 2M H_2SO_4. The third spectrum was taken after 900 minutes of operation. At this point the membrane would no longer transport copper.

---

be removed from a large volume of water. The minimization of organic carry-over in the treated water is essential. The chemistry summarized in eq. 4 was developed specifically for use in supported liquid membranes, and seems to work adequately in laboratory tests.[7,13] However the engineering development remains to be done, and, if past experience is a guide, this will bring to light needed chemical improvements as well.

Mechanism of Membrane Transport

Because of the limited area of supported liquid membranes it is critical that the details of the transport process should be understood, so that significant resistances can be minimized in so far as possible. Five stages can be identified in the transport process. These are: diffusion of the target ion through the semi-stagnant aqueous layer (Nernst layer) to reach the membrane; spontaneous reaction, with extraction, at the feed-membrane interface; diffusion of the target-carrier complex across the membrane under the impulse of a concentration gradient; spontaneous reaction, with liberation of the target ion and regeneration of the carrier, at the membrane-strip interface; and diffusion of the target ion, away from the membrane, across the Nernst layer, into the bulk of the stripping solution. The kinetics of the process can be simplified and straightforwardly analyzed provided that a steady state is reached. The condition for a steady state is that the change in the amount of target ion held in the membrane should be insignificant compared to the flux being measured. This can be met either by keeping the amount of target ion in the membrane small, or by keeping it relatively constant. If the condition is met, then the flux in the forward direction, in each of the five stages, J_n, where n identifies the stage, is given by a rate constant with units of cm s^{-1} times the concentration of the target ion in the feed. The k_n behave like conductances in

$$J_n = k_n C(fd) \tag{9}$$

series.[10,14] The inverse of each is called a resistance. The sum of these resistances is the inverse of an overall permeability, k, which is the observed first-order rate constant for depletion of the feed, k_{obs}, divided by the membrane area, a, and multiplied by the volume of the feed, v.[10,14] The total net flux in the forward direction, J is

$$\frac{1}{k} = \sum_n \frac{1}{k_n} \tag{10}$$

$$k = k_{obs}\ v/a \tag{11}$$

$$J = k_{obs} c(fd) \tag{12}$$

given by eq. 12, and the flux per unit area, j, is J/a. The smallest of the k_n have the largest effect on k, and it is on these that we should focus our attention if we wish to improve the performance of our membrane.

Equation 10 contains five unknowns — the five k_n. To obtain their magnitudes at least four subsidiary conditions must be found. This has been done in a variety of ways. If the overall transfer is spontaneous by at least a power of ten, as it always is in practical cases, and the hydrodynamics are made identical on the loading and stripping faces of the membrane, even if they are not well defined, $k_5 >> k_1$.[15] If k_5 is certain to be much larger than at least one of the other k_n it need not be further considered, especially since the aqueous hydrodynamic processes have little inherent interest. The membrane thickness can be varied by making multiple layer membranes.[15] If all other conditions are maintained constant, k_3 should be inversely proportional to the membrane thickness, which should not affect any of the other k_n.[15] Many investigators have noted that k_3 is proportional to the distribution coefficient of the target ion between the feed and the hydrophobic liquid phase. This distribution coefficient can be measured independently. Many have also made use of the fact that k_1 and/or k_5 can be increased until they have a negligible effect on k by increasing the flow rate of the aqueous solutions past the two faces of the membrane. It is also possible to construct an apparatus for which k_1 and/or k_5 can be calculated, by giving the membrane the configuration, and, therefore, the hydrodynamics, of a rotating disc.[14] A rather straightforward way to get information about k_4 is to make k_5 large, as described above, to hold k_1, k_2, and k_3 constant by using feed and membrane liquid of constant composition, and then to vary the composition of the strippant. Any variation in k which results can be attributed to a variation in k_4. As experience in the use of a particular type of apparatus builds up it should be possible to transfer k_1 from one experiment to another, since aqueous solution diffusion coefficients are not very structure-sensitive. If a multi-layered membrane is brought to a steady state, then quickly pulled apart and the target ion concentration of each layer separately determined, dc/dl, the concentration gradient of the target ion within the membrane, is directly available.[16] The result of such a study is shown in fig. 3.

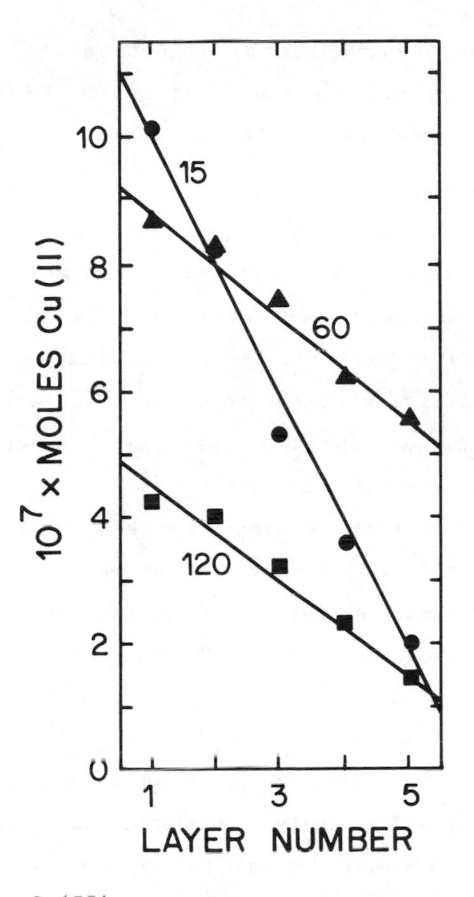

Figure 3. Cu(II) concentration profiles in multiple layer, solid supported liquid membrane for the selective transfer of copper from an acetic acid - acetate buffer to 2M H_2SO_4. The organic solvent is decalin. The copper carrier is LIX^R-64N, shown below eq. 2. The time, in minutes, at which each experiment was interrupted, is shown on the figure. Each layer was about 7 x 10^{-3} cm thick and the total thickness of the five-layer membrane was about 4 x 10^{-2} cm.

Since J can be measured, D/l, the effective diffusion coefficient in the organic phase divided by the membrane thickness, can be determined from Fick's first law, eq. 13.

$$j = (D/l)\ dc/dl \qquad (13)$$

The distribution coefficient of the target ion, times D/l, is k_3. This experiment also gives information about k_2 and k_4, by comparing the equilibrium concentration of target ion in the organic liquid directly adjacent to the feed and the strippant, respectively, with the concentrations actually found.[16] These techniques, and the apparatus in which they can be applied, have been recently reviewed.[17]

As an example of what these techniques can accomplish, they have shown that there are significant resistances at both the loading and stripping interfaces for the transfer of Cu^{2+} from SO_4^{2-} - HSO_4^- solutions of pH between 2 and 4, and 1-2 M H_2SO_4, using a phenolic oxime in a hydrocarbon solvent.[16,18] The resistance at the loading interface is greatly reduced when the feed contains an acetic acid - acetate buffer.[16] In sufficiently dilute solution k_2 depends on concentrations as shown in eq. 14, and saturation phenomena are observed at higher concentrations.[18]

$$k_2 = k'(\overline{RH})^2/(H^+) \qquad (14)$$

(k' is a concentration-independent parameter.)

A study[7] of nitrate transfer between near-neutral solutions and solutions with pH between 10 and 14 using the chemistry of eq. 4 reveals a small but significant resistance at the stripping interface. (Within the range of concentrations studied it never reduces k by more than two powers of 10.) Equation 15 shows the suggested dependence of k_4 on concentrations in the aqueous strippant.[7]

$$k_4 = C\{k_S K_{OH}(OH^-) + k_{OH} K_S(OH^-) / (NO_3^-) + k_{H_2O} K_S / (NO_3^-)\} \quad (15)$$

Equation 15 was derived from a suggested stripping mechanism.[7] The lower case k's are rate constants of individual steps of that mechanism and the capital K's are equilibrium constants. When the parameter, C, is multiplied by the NO_3^- concentration of the feed, it is believed to give the concentration of $Q^+NO_3^-{\cdot}HA$ at the stripping interface. With 3 adjustable parameters, eq. 15 successfully mimics 15 values of k. Rate equations of the complexity of 15 are unlikely to be unique, but it is derived from an intuitively satisfying model,[7] and will be helpful in designing and projecting the performance of devices.

It is common to point out the disagreements between investigators using these techniques to evaluate the various k_n,[17] and there is justification for doing so, but there is also a considerable measure of agreement about specific, well studied systems. Much of the apparent disagreement arises from studies using different conditions, different solvents, or entirely different chemical systems. In particular, no general evaluation of the importance of interfacial resistances is possible. Interfacial reactions, no doubt, show as much variety of mechanism, rate law, and rate constant as homogeneous solution reactions, and that is certainly considerable!

Limiting Flux

In preliminary evaluations of the possible applicability of a supported liquid membrane solution to a particular separation problem it is often useful to know what the most successful result could yield. To make such a calculation it was assumed that all resistances except those internal to the membrane could be removed. No suggestion for removing the internal resistance has yet come forward. The concentration of carrier-target complex at the loading interface was assumed to be 1M. The carriers and their complexes with the targets must be very water insoluble. All such materials, to date, have been of high molecular weight. Even if such a carrier is used without a solvent it seems unlikely that a concentration much over 1M can be achieved for the complex. For such large molecules in liquids which are generally fairly viscous it is hard to imagine diffusion coefficients larger than 10^{-6} cm^2 s^{-1}. A minimum membrane thickness of 2.5×10^{-3} cm has been used. This is the thickness of the thinest currently available porous plastic material. However, to get the effective diffusion coefficient from the real one, one must divide by a tortuousity factor and multiply by the fractional void volume. These modifications make the effective diffusion coefficient about a factor of 10 smaller than the real one.[15] Thus the value used is optimistic. Even with these best-case estimates the calculated flux is only 4×10^{-7} moles cm^{-2} s^{-1}. Thus it would seem unwise to embark on a supported liquid membrane research program the success of which would depend on achieving a flux exceeding or even closely approaching this value.

Conclusion

It seems likely that supported liquid membranes will soon find a place among large scale separation processes.

The most likely candidates are separations in which a small amount of target ion must be removed from a large volume of aqueous solution, either because it is valuable, or because its presence degrades the value of the stream. Because the success of a supported liquid membrane separation depends on a number of factors interacting in a complicated way, engineering development can benefit a great deal from basic studies, which can now be rationally organized.

References

[1]D.W. Agers, J.E. House, R.R. Swanson and I.L. Drobnik, Trans. Soc. Mining Eng., 1966, 191.

[2]Anon., Chem. Eng. News, 1956, 34, 2590.

[3]W.A. Rickelton, D.S. Flett and D.W. West, Proc. ISEC '83, Denver, Colorado, U.S.A., Aug. 1983, p. 189.

[4]Anon., Chem. Br., 1984, 20, 194.

[5]T. Mattila and A. Lokio, Environ. Pollut. Manage., 1979, 9, 68.

[6]M.M. Kreevoy and C.I. Nitsche, Environmental Sci. and Technol., 1982, 16, 635.

[7]M.M. Kreevoy, A.T. Kotchevar and C.W. Aften, Separation Sci. and Tech., in press.

[8]P.F. Thelander, L.A. Hasledalen and M.M. Kreevoy, J. Chem. Ed., 1980, 57, 509.

[9]K. Sollner and G.M. Shean, J. Am. Chem. Soc., 1964, 86, 1901.

[10]E.L. Cussler, AIChE J., 1971, 17, 1300.

[11]W.C. Babcock, D.J. Kelly and D.T. Friesen, Proc. ISEC '83, Denver, Colorado, U.S.A., Aug., 1983, p. 373.

[12]Estimated from technical data supplied by Bend Research, Inc.

[13]M.M. Kreevoy and A.T. Kotchevar, U.S. Patent applied for.

[14]W.J. Albery, J.F. Burke, E.B. Leffler and J. Hadgraft, J. Chem. Soc. Faraday Trans. 1, 1976, 72, 1618.

[15]L.A. Ulrick, K.D. Lokkesmoe and M.M. Kreevoy, J. Phys. Chem., 1982, 86, 3651.

[16]S. Djugumović, M.M. Kreevoy and T. Škerlak, J. Phys. Chem., 1985, 89, 3151.

[17]G.J. Hanna and R.D. Noble, Chem. Revs., 1985, 85, 583.

[18]W.J. Albery, R.A. Choudhery and P.R. Fisk, Faraday Discuss. Chem. Soc., 1984, 77, 53.

Removal of Cadmium Contained in Industrial Phosphoric Acid Using the Ionic Flotation Technique

By E. Jdid,[1] P. Blazy,*[2] J. Bessiere,[3] and R. Durand[4]

[1]ECOLE NATIONALE DE L'INDUSTRIE MINÉRALE, BP 753, AGDAL-RABAT, MOROCCO

[2]CENTRE DE RECHERCHE SUR LA VALORISATION DES MINERAIS, BP 40, 54501 VANDOEUVRE CEDEX, FRANCE

[3]LABORATOIRE DE CHIMIE ET ELECTROCHIMIE ANALYTIQUE, UNIVERSITÉ NANCY I, BP 239, 54506 VANDOEUVRE CEDEX, FRANCE

[4]URANIUM PECHINEY, DIRECTION RECHERCHE ET DÉVELOPPEMENT, TOUR MANHATTAN CEDEX 21, 92087 PARIS LA DEFENSE, FRANCE

1. Introduction

The problem which cadmium in phosphates poses is directly linked to the utilisation of phosphate fertilizers and it may be considered specific to industrialised countries. Furthermore, this problem is due especially to the cadmium accumulation in the soil over the years. The quantities of phosphates fertilizers used and the cadmium which is given off annually by arable land hectare are shown in table 1 (1).

Table 1 - Cadmium accumulation into the soil of EEC countries, introduced by phosphate fertilizers.

COUNTRY	kg P_2O_5/ha arable land/year	Cd g/ha arable land/year
Italy	29,80	1,6
Netherlands	57,00	4,6
United Kingdom	64,00	6,5
West Germany	96,00	4,6
Belgium	114,00	9,4
France	76,00	5,4
Denmark	46,00	2,4
Eire	124,00	6,6

Elsewhere, the quantity of cadmium carried by the phosphate fertilizers depends especially upon the phosphate's origin.

Table 2 - Average Cd contents of phosphate rocks according to country of origin

COUNTRY	Cadmium (Cd) g/t
Algeria	18
Israel	10
Jordan	10
Morocco	18
Senegal	76
Togo	48
Tunisia	18
United States	
. Florida, North Carolina	10
. Wyoming	45
Soviet Union (Kola)	0,3

The only industrial cadmium elimination process takes place on the phosphate calcination at 1150° C (1). It has been used on the Nauru Island phosphates and at the same time increases the reactivity of these phosphates and lowers their cadmium content from 80 ppm to 20 ppm. However, this process is invalid for phosphates rich in aluminium, silica and iron oxides. Tests carried out on southern United States phosphates which have strong contents rich in these elements, failed and gave an agglomerate of smelted rocks.

Other very recent European tentatives (dating from the latter 1970's) have dealt with cadmium's elimination starting with phosphoric acid obtained through acidic treatment. This allowed the development of processes based upon the following techniques :

1) Cadmium precipitation by H_2S under pressure after phosphoric acid dilution at 13-20 % P_2O_5 and its neutralization at pH = 0,9-1,5 (2)
2) Solvent extraction, using either aminosalts (such as tridodecylamine-HCl) which give extraction efficiencies increasing with acid concentration (3), either dialkylesterdithiophosphoric acids to extract the cadmium from the neutralized acid at pH = 0,5-1 (1) or concentrated acid (5). In the latter case, organic matters produce micro-emulsions and the phase separation is slow.

3) The separation by fixation of a dithiophosphoric ester impregnated upon porous support : the support may be active carbon, perlite, kieselguhr, carbon black (obtained by the pyrolysis of the acetylene or gazification under oxygen pressure of heavy oils), silicates or amino-silicates, in particular the zeolites (6,7).
4) The ionic flotation upon which this communication focuses (8, 9, 10, 11, 12).

2. Presentation of the ionic flotation technique

In a simple analysis, a surface active-ion (collector) is introduocd into the solution, in which it forms a complex -either soluble or in the form of a precipitate- (sublate) with the ion that is to be floated (colligend). The sublate adsorbs on the gas bubbles, which rise through the solution to the surface, and collects in a froth from which it can be recovered as a solid (13, 14, 15). One must remember that one of the ionic flotation advantages, in the case of a precipitated sublate, is to reduce considerably the liquid-solid volume to filter.

This technique was especially developed at bench laboratory scale work in aqueous solutions (pH = 1-11) and has not yet been applied to the industry. Its application to the recovery and/or removal of metallic cations from industrial phosphoric acid has encountered two major problems :

1) The choice of the collector : the latter must be stable and capable of giving a complex with the ion to be floated, directly in the phosphoric acid concentrated at 30 % P_2O_5 (H_3PO_4 5,5 M ; Ro(H) =-1,9).
2) The bubbles' production : the H_3PO_4 solution being relatively viscous, it is necessary to find a means of generating bubbles without having to dilute it and consequently to modify it for a subsequent process.

3. Ionic flotation of cadmium in phosphoric acid solutions

The collector selected to float the cadmium contained in H_3PO_4 is the sodium diethyldithiophosphate known as "Hostaflot LET" commercialized by the society Hoechst. Its formula is :

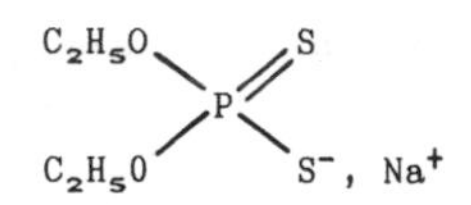

In H_3PO_4 5,5 M ($\approx$ 30 % P_2O_5), the LET collector gives with Cd^{2+} a precipitate containing 22,33% Cd, a value close to the 23,30% theoretical content, corresponding to the Cd $(LET)_2$ compound of stoichiometry 1:2. The pK_S reaction is equal to 11,7.

First, we will consider the cadmium elimination alone from synthetic H_3PO_4 solutions and then the ionic flotation of this element directly in the industrial phosphoric acid or in 5,5 M H_3PO_4 synthetic solutions containing species susceptible of interfering. The ionic flotation apparatus used in the laboratory for this study are shown in figure 1.

3-1. Cadmium ionic flotation alone in a synthetic phosphoric acid solution.

Practically speaking, the cadmium is introduced into the solution in nitrate form and the LET collector in concentrated aqueous solution form (Na-LET 2,7 M) for a conditioning time of 20 mn. The flotation operation follows for a 10 mn period, the gas (air, nitrogen) flow rate being 2 l/h. The solution captured by froth is under these conditions inferior to 5 %. The flotation efficiency is measured by the polargraphic analysis in the DMSO and in the phosphoric solutions. As the cadmium precipitates within the solution itself and the precipitate formed is well flocculated, the flotation efficiency is directly related to the metallic element precipitation ratio.

We have successfully examined the following parameters (table 3 and figure 2) :

a) the collector concentration, or Φ = [LET] M / $[Cd^{2+}]$ M.

b) the Cd^{2+} ions concentration.

c) the H_3PO_4 acid concentration.

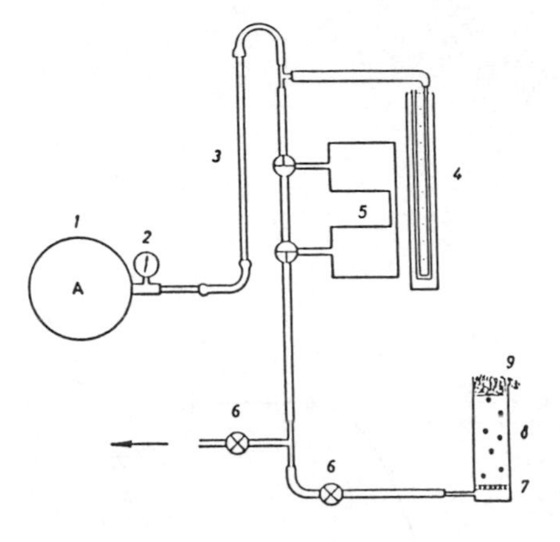

Figure 1 : Schematic drawing of flotation apparatus

1 - Nitrogen bottle, 2 - Micro-pressure reducer, 3 - Flow meter, 4 - Manometer, 5 - Expansion chamber, 6 - Control taps, 7 - Porous glass frit, 8 - Flotation cell, 9 Froth.

Table 3 - Effect of Cd and LET concentrations on ionic flotation of cadmium in 5,5 M synthetic H_3PO_4

$[Cd^{2+}]$ M	$\phi = \frac{[LET]}{[Cd]}$	Removal %
$5 . 10^{-4}$	1	33
$5 . 10^{-4}$	2	74
$5 . 10^{-4}$	4	95
$2 . 10^{-4}$	4	81
$5 . 10^{-4}$	4	96
$1 . 10^{-3}$	4	96

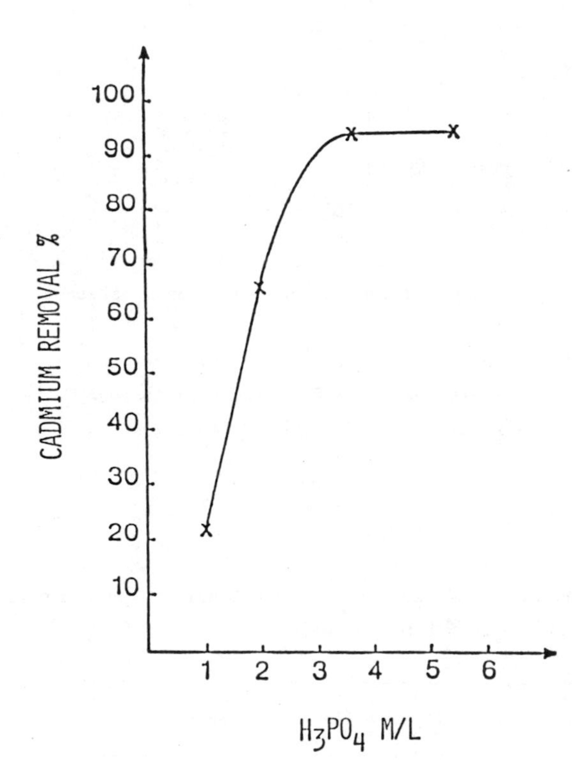

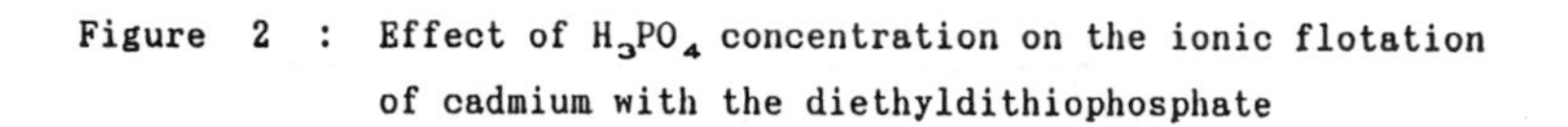
Figure 2 : Effect of H_3PO_4 concentration on the ionic flotation of cadmium with the diethyldithiophosphate

$[Cd^{2+}] = 5\ 10^{-4}$ M ; $\Phi = 4$

These results may easily be interpreted, using the classical precipitation rules :

a) The flotation efficiency increases with the Φ value as a result of the mass effect. The difference observed between the 74 % removal obtained for Cd 5 10^{-4} M with a value of $\Phi = 2$ and that of the precipitation rate (86 %) is due to the residual adsorption of the floated precipitate on the flotation cell's walls. Note must be taken that all the precipitate formed floats and that the solution is perfectly limpid after the flotation operation.

b) The flotation removal increases for the same Φ value, with the initial cadmium concentration. This result corroborates the fact that the higher the initial species concentration is, the more the reaction of precipitation is quantitative. The removal R obeys the relation :

$(1 - R)\ (\Phi - 2\ R)^2 = K_S/C_0{}^3$

c) Finally, a better ionic flotation efficiency may be observed when the acid concentration increases. It is important to note that the LET collector is capable of floating the cadmium in a H_3PO_4 11,5 M solution, even though it precipitates in acid form in the solution. However, the flotation operation is longer because of the solution's high viscosity.

The general results confirm the fact that the cadmium may be floated in an entirely satisfactory way, in concentrated phosphoric acid solutions. The collector's excess remains soluble in the 5,5 M H_3PO_4 solution. Elsewhere the flotation efficiency is temperature independent.

<u>3-2. Cadmium ionic flotation in industrial phosphoric acid.</u> Studies were carried out essentially on the phosphoric acid resulting from the sulfuric attack of the TAIBA phosphates (Senegal). This black acid containing organic matters has a content of 28 % P_2O_5 (H_3PO_4 5 M) ; 74 mg/l Cd (6,5 10^{-4} M) ; 46 mg/l Cu (7,2 10^{-4} M) and 7 g/l Fe (1,25 10^{-1} M). This phosphoric acid is considered as highly cadmiferous. Furthermore, polarographic tests show that iron exists in the trivalent state.

The first flotation tests carried out on these industrial solutions gave mediocre results. The LET's addition ($\Phi = 4$, versus cadmium concentration) involves the appearance of a thick yellowish precipitate

which floats with difficulty. The results are barely reproducible, the removals varying from 5 to 50 %. The solutions remain equally disturbed.
In the case where cadmium floats conveniently in a synthetic solution and where the liquid, after flotation, is completely clarified, this difference may only be explained by the existence of secondary reactions which noticeably disturb the Cd^{2+} species flotation process. With regard to the LET's redox properties, one may anticipate a collector consumption by species which are easier to complex than cadmium or a degradation of the collecting properties of diethyldithiophosphate because of its oxidation by the strongly concentrated ferric ions.
A certain number of orientation tests upon industrial solutions enabled us to show the major influence of 3 parameters which anyway are related : the solutions' potential, the presence of copper and the temperature.
In order to facilitate the presentation of the results, the examination of these different parameters was made simultaneously on the synthetic and industrial H_3PO_4 solutions.
The Cd and Cu were analysed by atomic absorption.

3-2.1 Effect of copper

3-2.1.1 Interaction Cu^{2+} - LET. $Cu^{2+}$and Cd^{2+} ions are present in the TAIBA phosphoric acid at about the same molar concentration. It is known that copper ions generally possess a strong affinity for the organo-sulfuric compounds, in particular the thiol ones. So it is important to ascertain if they are more reactive than Cd^{2+} ions towards the collector LET.
The polarographic plottings (figure 3) carried out on a synthetic solution, H_3PO_4 5,5 M, containing equal concentrations of Cd^{2+} and Cu^{2+} ions, show unambiguously that the copper reacts quantitatively first. Cadmium begins to precipitate after total copper consumption.
The first conclusion which can be drawn is that the presence of copper in the industrial solutions leads to a noticeable collector consumption. If the chemical reaction, Cu^{2+}-LET, is examined more closely, we note the intervention of two processes at the same time, complexation and oxido-reduction. The same case is encountered when the Cu(II) action is envisaged on the iodide or on the xanthates (20,21).

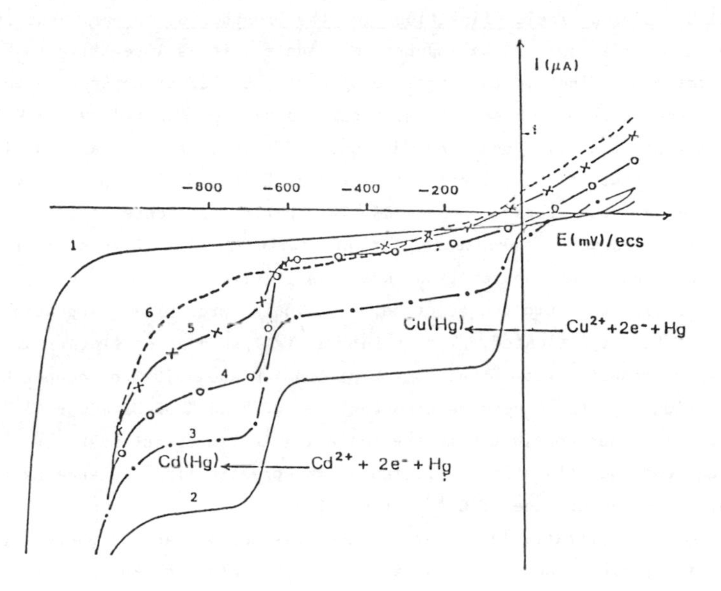

Figure 3 : Precipitation of cadmium with sodium diethyldithiophosphate (LET) in the presence of copper in 5,5 M synthetic H_3PO_4

1. 5,5 M synthetic H_3PO_4
2. $[Cd^{2+}] = [Cu^{2+}] = 5\ 10^{-4}$ M
3. (2) + 5,0 10^{-4} M LET
4. (2) + 1,0 10^{-3} M LET
5. (2) + 1,5 10^{-3} M LET
6. (2) + 2,0 10^{-3} M LET

The reaction is the following :

$$Cu^{2+} + 2\ (C_2H_5O)_2PS_2^- \rightleftharpoons (C_2H_5O)_2PS_2Cu(\downarrow)^* + 1/2\ [(C_2H_5O)_2PS_2]\ (\downarrow)$$

$(\downarrow)^*$: precipitate

which is symbolized :

$$Cu^{2+} + 2\ LET \rightleftharpoons CuLET\ (\downarrow) + 1/2\ (LET)_2\ (\downarrow)$$

The result of the LET action on Cu^{2+} is the appearance of a CuLET(↓) precipitate and a solution disturbance due to the formation of the oxidation product $(LET)_2(\downarrow)$.

3-2.1.2 <u>Cadmium ionic flotation in the presence of copper</u>. Reactions between Cu(II) and LET having been elucidated, it is imperative to know if cadmium may float in the presence of Cu^{2+}, the LET concentration being of course suitably chosen. So many tests have been carried out on H_3PO_4 5,5 M synthetic solutions. The Cd^{2+} and Cu^{2+} concentrations are equal to 5 10^{-4} M, that of the collector is fixed at 2 10^{-3} M. It implies a value of $\Phi = 4$ with regard to cadmium. This ratio represents precisely the theoretical quantity necessary for precipitating the 2 ions, if it is supposed that these two reactions are quantitative.

After adding LET into the solution, a yellowish precipitate appears that may be totally floated. The solution is limpid after flotation. The analysis results show that 73,2 % of cadmium and 99,6 % of copper have been floated, the concentrate content being 12,38 % of Cd and 9,43 % of Cu. One may thus conclude with the following two important results :

1) The cadmium flotation efficiency is practically the same as that obtained in the absence of Cu^{2+} for $\Phi = 2$.

2) Copper quantitatively floats in the presence of cadmium whereas this is not possible when it is alone in the solution. Furthermore, the solution's limpidity implies the presence of the $(LET)_2$ in the concentrate. This original property is worth studying in detail for it may be applied, in certain cases, to the elimination of ions which give precipitates difficult to float.

We may now conclude from the above that when iron (III) is absent, cadmium is able to float even in the presence of copper (II) so long as a sufficient concentration of the collector is used. Copper is equally floated.

3-2.2 <u>Iron effect</u>

3-2.2.1 <u>Iron (III) concentration</u>. The thermodynamic calculations have demonstrated that the LET may be oxidized by ferric ions if the latter are sufficiently concentrated. Note should be taken that such calculations are not accounted for eventual complexations by the LET of the Fe^{3+} and Fe^{2+} ions as it is the case for the xanthates (21).

So far, in 5,5 M synthetic H_3PO_4 solution, we have determined the flotation efficiency of cadmium contained at 5 10^{-4} M under the following conditions : first alone in the presence of a variable iron (III) concentration, then by taking a value of $\Phi = 4$ (versus Cd^{2+} concentration) and finally in the presence of 5 10^{-4} M Cu(II) and 10^{-1} M Fe(III), the Φ value being variable. These tests were carried out at ambient temperature (tables 4 and 5).

Table 4 - Effect of iron concentration on the ionic flotation of cadmium in 5,5 M synthetic H_3PO_4. $\Phi = 4$; T = 25° C

[Fe(III)] M	Cd removal %	Cd concentrate content %
-	99,14	22,83
5 10^{-3}	98,93	22,24
1 10^{-2}	99,92	22,13
5 10^{-2}	96,10	22,48
1 10^{-1}	81,92	21,98

Table 5 - Collector concentration effect (i.e. Φ) on the ionic flotation of cadmium in 5,5 M synthetic H_3PO_4 ; 5 10^{-4} M Cd(II) ; 10^{-1} M Fe(III) ; 5 10^{-4} M Cu(II) ; T = 25° C

Φ	Removal %		Concentrate content	
	Cd	Cu	Cd %	Cu %
4	49,87	89,79	10,07	9,59
6	96,04	96,27	12,57	7,35
8	98,69	96,27	12,60	6,94

For the cadmium alone, the removal decreases when Fe(III) concentration is higher than 5 10^{-2} M. Consenquently a collector consumption occurs by the ferric species, either by oxidization or by complexation. Elsewhere, the fact that the precipitation is accompanied by a slight solution disturbance, is in favour of an oxidation phenomenon.

For a value of Φ at least equal to 6 and in the presence of copper, a good cadmium removal is obtained although the iron (III) concentration remains high. The cadmium content of the concentrates logically increases with the value of Φ and the copper removal is also very satisfactory. Then, if the results are compared according to the presence or not of Fe(III) ions, it can be shown that the fall of the cadmium flotation efficiency is about 25 % for Φ = 4. In fact, the removal decreases from 73,2 % to 49,8 %.

Accordingly, the tests carried out in synthetic solutions of phosphoric acid, at ambient temperature, demonstrate that the presence of Fe(III) in a high concentration (10^{-1} M) does not inhibit the cadmium flotation in the presence of copper, providing that the collector concentration is increased. But the use of an excess of collector can be an inconvenience in the way it may pollute the acid because of its solubility in H_3PO_4. This excess may also react on ions and give precipitates which float with difficulty or disturb the cadmium flotation (co-adsorption). In order to avoid such inconveniences and so as to float the cadmium in the desired conditions, it is possible to envisage a chemical reduction of the medium by using iron powder. As we shall see later on, the temperature effect compels phosphoric acid reduction for the flotation of cadmium.

3-2.2.2 <u>Effect of the reduction on the ionic flotation of cadmium in the presence of iron and copper.</u> The iron powder is a powerful reducing agent in phosphoric solution. It allows the reduction of cupric and ferric ions. The copper cementation upon metallic iron is easily observable, the latter changing its colour. Likewise, it is possible to reduce the solvent with hydrogen releasing. The reactivity order of metal iron on the ions and on the solvent depends obviously on the solutions' agitation.

Because the medium reduction is an operation which may modify the iron concentration of the solutions, it is essential to know if a complete or a partial acid reduction is necessary for the ionic flotation of cadmium. This reduction can be quantified in two ways : first, by the reduction time and secondly by the Fe(II) / Fe(III) ratio which lay down the

solution potential because of its high iron content. The reduction is a parameter dependent noticeably on the agitation, on the divided set of the iron powder and of its superficial degree of oxidation. The comparison between two mountings is very difficult. However, it is more rigorous to evaluate, for any reduction time, the Fe(II) / Fe(III) ratio. This is easily accomplished by polarographic plots in a test intake of the solution diluted in pure 5,5 M synthetic H_3PO_4.
For synthetic solutions containing initially 5 10^{-4} M Cd(II), 5 10^{-4} M Cu(II) and 10^{-1} M Fe(III), the results (table 6) confirm that the cadmium flotation efficiency increases with the acid's reduction ratio. With a Fe(II) / Fe(III) ratio of an approximate value of 2, it is possible to achieve the maximum removal of cadmium, which however remains slightly lower to that acquired in the case of solutions free of iron and for the same value of Φ = 4 (86 % instead of 95 %). This difference may be explained partly by the presence of copper which has not been entirely reduced (eliminated by cementation).

Table 6 - Acid reducing effect on the ionic flotation of cadmium contained in 5,5 M synthetic H_3PO_4
Cd(II) = 10^{-4} M ; Cu(II) = 5 10^{-4} M ; Fe(III) = 10^{-1} M
Φ = 4 ; T = 25° C

Fe (II) / Fe(III)	Residual Cu after reduction			Cd removal %	Concentrate content	
	mg/l	M/l	%		Cd %	Cu %
-	31,8	5,0 10^{-4}	100,00	48,87	10,07	9,59
0,3	21,8	3,4 10^{-4}	70,32	68,85	13,39	8,61
0,6	14,0	2,2 10^{-4}	45,16	82,15	16,64	5,56
2,0	3,9	6,1 10^{-5}	12,58	86,05	20,39	2,06
3,7	3,3	5,2 10^{-5}	10,65	85,27	20,73	1,86

3-2.3 Temperature effect. The results mentioned previously deal with operations carried out at ambient temperature (≈ 25° C). Some qualitative tests made on synthetic solutions show that this parameter has a great influence on the cadmium flotation efficiency in the presence of iron (III). The oxidation of the collector by ferric ions in 10^{-1} M concentration is more noticeable at 60° C than at 25° C. This fact implies an important cloudiness of the solution. Likewise, the cadmium precipitation from a solution containing Fe(III) seems to be partially less quantitative, the dithiophosphate ions being consumed by the ferric ions.

The ionic flotation of cadmium by LET in the presence of Fe(III) is thereby considerably endangered when taken place at temperatures greater than 40° C. This leads to use a very high value for Φ, resulting in significant acid pollution with an inadequate efficiency. But this important inconvenience is avoided if the phosphoric acid solution is previously reduced. In fact, it has been reported that the flotation of cadmium alone by LET in 5,5 M H_3PO_4 is little affected by a temperature variation. Qualitative tests realised on synthetic solutions show that after Fe(III) reduction, the cadmium precipitation by the LET, at about 50° C, provide results very close to those obtained at ordinary temperature.

The temperature effect on the ionic flotation of cadmium will be reconsidered in detail in the case of wet process industrial phosphoric acid of TAIBA (Senegal).

3-2.4 <u>Ionic flotation of cadmium in the TAIBA phosphoric acid</u>. The three parameters which govern the removal of cadmium from phosphoric acid solutions will be examined. They are the temperature, the potential and the copper effect.

3-2.4.1 <u>Temperature effect</u>. Accordingly to the precedent studies, the temperature parameter alone governs the treatment process of the phosphoric acid. If industrial operations of ionic flotation take place at relatively high temperatures, it is largely probable that an acid reduction will be required. But therefore the influence of this parameter must be measured with a great precision.

Results dealing with cadmium removal from the non reduced TAIBA phosphoric acid are compiled in table 7. The value of Φ has been fixed at 6 versus cadmium concentration.

Table 7 - Temperature effect on the cadmium ionic flotation. H_3PO_4 TAIBA

Temperature (°C)	21	33	43	53
Cd removal %	87,16	65,05	27,69	00,00

Temperature variation from 21 to 43° C makes the cadmium flotation efficiency decrease from 87,16 to 27,69 %. At 53°C, cadmium stops floating. The temperature effect is therefore determinant.

3-2.4.2 Effect of potential and of copper concentration. It is necessary to know the reduction ratio of the acid in order to obtain a satisfactory flotation efficiency. The evolution of cadmium removal has been studied according to the Fe(II) / Fe(III) concentrations ratio which corresponds to fixed reduction times. The tests have been carried out first at ambient temperature (table 8), and secondly at variable temperature (table 9).

Table 8 - Effect of the reduction ratio [Fe(II) / Fe(III)] on the cadmium ionic flotation. TAIBA H_3PO_4 ; Φ = 4 ; T = 25° C

Fe(II) / Fe(III)	-	0,5	2,0	3,9	8,5
Cd removal %	48,35	88,20	93,52	95,07	97,18

In table 8 it appears that the efficiency reached is already quite satisfactory with a Fe(II) / Fe(III) ratio equal to 2. It is not necessary to reduce strongly the acid when the phosphoric acid solution is at about 25° C. However, it is obvious that at a higher temperature the reduction ratio must be higher as shown in table 9. This table gives the whole results which show the evolution of the cadmium flotation efficiency in the TAIBA phosphoric acid, according to the temperature and to the acid reduction ratio. The copper behavior and the cadmium residual content are also mentioned.

Table 9 - Effect of the reduction ratio and of the temperature on the ionic flotation of cadmium. TAIBA H_3PO_4 ; $\Phi = 4$

Temperatures °C	$\frac{Fe(II)}{Fe(III)}$	Residual content (mg/l)			Removal %	
		Cu		Cd	Cu	Cd
		after reduction	after precipitation with LET			
20	-	46,00	1,20	38,30	97,39	48,35
	2,7	12,55	0,50	2,10	98,91	97,17
	15,0	2,90	0,10	0,80	97,78	98,92
40	- *	46,00*	2,10*	53,80*	95,43*	27,69*
	5,0	21,38	0,60	6,05	98,68	91,82
	14,3	2,20	2,13	0,85	95,37	98,85
	∞	1,00	0,10	0,80	99,78	98,92
50	- *	46,00*	2,00*	74,4*	95,65*	00,00*
	12,4	1,50	0,10	2,00	99,78	97,30
	∞	1,20	0,12	1,00	99,74	98,65

* These values correspond to $\Phi = 6$

So it appears that at 40 or 50° C, the cadmium flotation is entirely satisfactory, once the solution has been reduced. Thereby it is possible to admit that a Fe(II) / Fe(III) ratio of about 4 to 6 is sufficient to float the cadmium under good conditions when the temperature is 40°C. Elsewhere, a temperature increase accelerates the reduction kinetic and the cementation kinetic of copper, which enables a bigger availability of the oollcctor fur the Cd^{2+} ions. After the flotation, solutions are limpid. The floated concentrate titrates about 18 % Cd, which is a very good selectivity in the reduction medium.

4 - Industrial execution of the ionic flotation technique in 30 % P_2O_5 phosphoric acid solution

The industrial phosphoric acid is a more valuable product than metallic cations which it contains. Therefore, it is necessary to operate the removal or the recovery of these cations by the ionic flotation technique taking care not to lose too much acid.

Laboratory tests realised with a porous bottom cell have shown that the phosphoric acid content of the froth can be lower to 5 %, but according to our knowledge, this bubble method has, at the industrial scale, the disadvantage of clogging the pores. Then the other suitable possibilites are air pressurized apparatus used for water processing installations and turbine flotation cells used for mineral flotation. The first ones were early found inappropriate probably because of the viscosity of the phosphoric acid solution. The bubbles obtained through an expansion after a 7 bars pressurization are larger (≈ 100 μ diameter) than in the water (≈ 40 μ diameter) and too much dispersed the ones from the others. This does not create a sufficient volume to ensure the carriage towards the surface of the totality of the "collector-cation" hydrophobe precipitate. The second type of apparatus, that's to say the turbine flotation machines, provide excellent floating conditions in the concentrated phosphoric acid but it presents the disadvantage to provoke a too much important liquid drive in the froth which is estimated to 20 - 30 % with a Wemco laboratory cell for the flotation of the $Cd(LET)_2$ precipitate. It is consequently necessary to look for another way of generating the bubbles to float the "collector-metallic cations" precipitate which is present in the 30 % P_2O_5 phosphoric acid and may enable:

- a good solid-liquid separation efficiency ;
- a weak acid content in the froth with the creation of a sufficient air volume ;
- a reasonable execution time for these operations.

So a new type of equipment has been tested, initially imagined to carry out any gas transfer in a liquid (11, 21), (oxygenation, ozonation, gas washing...). This concerns the gas-liquid contactor "IMOX" which results from common work between IRCHA and MOTEUR MODERNE.

4-1 <u>Cadmium ionic flotation in synthetic H_3PO_4</u> According to the mounting described in figure 4, we have realised tests of ionic flotation of cadmium from 6 M H_3PO_4 synthetic phosphoric acid. The procedure conditions are the following :

- 40 to 50 liters synthetic 6 M H_3PO_4 charged with Cd ;
- Collector : Sodium diethyldithiophosphate,
  Φ variable : 2,5 to 4 ;
- Precipitation time : 15 to 20 mn ;
- Phosphoric acid flowrate : 450 l/h ;
- Air flowrate (natural aspiration by IMOX) : 100 - 110 l/h ;
- Output velocity of H_3PO_4 from IMOX : 1,8 m/s

The results obtained are reported in table 10. They show the necessity to recycle one or two times the phosphoric acid solution containing the $Cd(LET)_2$. A single passage in the contactor is not sufficient to gather the whole precipitate at the solution's surface. This recycling in closed-circuit allows the decrease of the precipitate quantity in the phosphoric acid to less than 5 mg/l in 5 minutes. This gives a good purification of the precipitate and an excellent cadmium removal rate, close to about 90 %.

It is also important to note that the passage in the contactor of the solution containing the precipitate does not alter the floatability properties of the latter.

Elsewhere, the bad precipitation rate observed with $\Phi \approx 2,5$ is due to a bad homogenization of the phosphoric acid solution (agitation not adapted to such an important volume of acid : 40 - 50 l). We may see too, that the best results are obtained with the 25 l cell which allows the longest retention time.

As a conclusion we note that the IMOX gas-liquid contactor is well adapted to the ionic flotation in 30 % P_2O_5 phosphoric acid and that the encouraging results, obtained on synthetic acid, leaa us to realise further tests at pilote scale on industrial phosphoric acid.

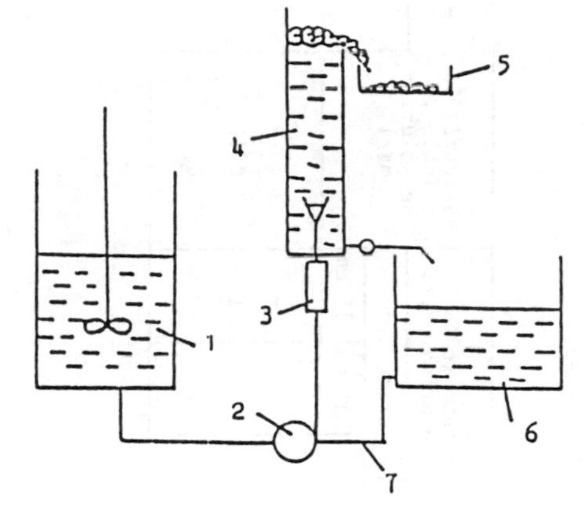

Figure 4 - Ionic flotation : experimental mounting

1. Flocculation tank
2. Centrifugal pump
3. Gas-liquid contactor (IMOX), 0,5 m^3/h (liquid flowrate)
4. Cylindrical flotation cell
5. Container for the flook recovery
6. Tank for the purified solution
7. Recycling circuit

Table 10 - Ionic flotation of cadmium in 6 M H_3PO_4

IMOX gas-liquid contactor

Cells	Initial content of the solution Cd mg/l	Φ	Residual precipitate in the solution after flotation, mg/l				Residual Cd soluble in the solution mg/l	Cadmium removal % after	
			1 passage	Recycling time				1 passage	5 mn recycling time
				5 mn	10 mn	15 mn			
6,5 l	62,6	2,5	35	-	-	-	36,9	≈ 30	-
6,5 l	58,1	4	24	3	2	1	0,42	≈ 90	98
25 l	57,0	4	19	< 1	-	-	0,17	≈ 92	> 99

4-3 <u>Pilot tests of ionic flotation in 30 % P_2O_5 industrial phosphoric acid.</u> The society Unitec-Wemco has at disposal a flotation pilot which uses a bubble production system, similar to the IMOX, and called "Nozzle air" Venturi process. This apparatus is equiped with four flotation cells and with an exhausting box, all of them disposed in series and with a unit capacity of about 30 l. This pilot scheme is given by the figure 5. The Unitec-Wemco pilot, equipped with the gas-liquid contactors of "Venturi" type can be used for ionic flotation but in comparaison to the IMOX, it has the disadvantage to cause a great turbulence at the solution's surface and to give a rather important air/volume ratio (2,5 times the IMOX one). This provokes unfavourable reoxidation phenomena in the reduction medium. So we have then equiped this pilot with two IMOX gas-liquid contactors, fitted on the first two cells. The operating conditions are the following :

- phosphoric acid volume : 200 l ;
- H_3PO_4 feed flowrate : 530 - 690 l/h ;
- air flowrate (natural aspiration) : 175 - 220 l/h ;
- feed pressure : 0,7 to 1,5 bars.

The industrial phosphoric acid used to achieve the tests is a mixture of 65 % from Togo and 35 % from Morocco. It titrates 28 % P_2O_5 in weight, 7 g/l iron and 41,5 mg/l cadmium. Two tests have been executed and their results are reported in table 11. They confirm the efficiency of the IMOX gas-liquid contactors for the ionic flotation in 30 % P_2O_5 industrial phosphoric acid. With only two cells among the four ones of the Wemco pilot it is possible to float more than 90 % of the precipitate formed in the phosphoric acid solution. These results agree with those obtained at laboratory scale on small volumes, even though the reduction was sometimes difficult to be realised. The liquid loss was included between 5 and 10 % for all the tests.

5 - <u>Fore project of industrial realisation</u>

We suppose a phosphoric acid production unit of 300 000 t/year of P_2O_5 which should treat a phosphate relatively charged in cadmium. Then the raw acid properties would be approximately the following :

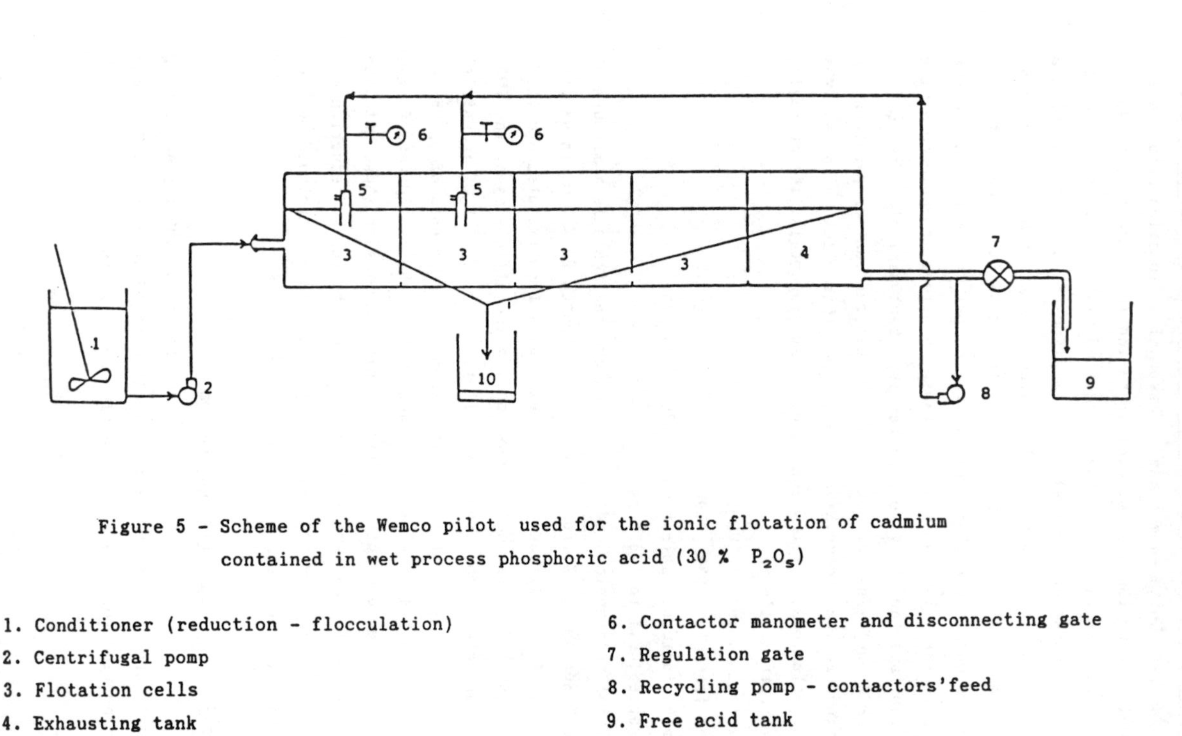

Figure 5 - Scheme of the Wemco pilot used for the ionic flotation of cadmium contained in wet process phosphoric acid (30 % P_2O_5)

1. Conditioner (reduction - flocculation)
2. Centrifugal pomp
3. Flotation cells
4. Exhausting tank
5. "IMOX" gas-liquid contactor
6. Contactor manometer and disconnecting gate
7. Regulation gate
8. Recycling pomp - contactors'feed
9. Free acid tank
10. Concentrate flotation tank

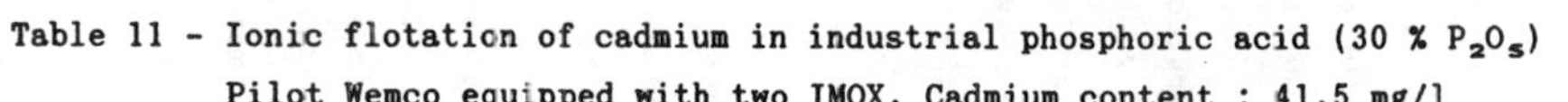

Table 11 - Ionic flotation of cadmium in industrial phosphoric acid (30 % P_2O_5)

Pilot Wemco equipped with two IMOX. Cadmium content : 41,5 mg/l

Test n°	Residual content in solution mg/l	Φ	Precipitation rate %	Solid in solution before flotation mg/l	Residual solid after flotation	Weight recovery of the precipitate flotation %
1	1,2	4	97,21	125	14	88,80
2	< 1	4	97,50	190	17	91,05

- flowrate : 100 m^3/h of 29 % P_2O_5 phosphoric acid at 60° C ;
- Contents : .Cd : 70 mg/l
  .Fe^{3+} 7 g/l
  .Fe^{2+} negligible

The scheme of the ionic flotation circuit to be shunt in the phoshoric acid is given by figure 6. This scheme only uses classical material. The phosphoric acid plant reduction by means of iron is now well controlled for it is usual in the uranium recovery of phosphoric acid by OPPA way (solvent extraction). The investment cost of such a plant, for a 100 m^3/h phosphoric acid flowrate, would be of about 20 M FF, according to the present European rules.

The global operating cost including the reagents, services, labor, maintenance, amortization and financial costs would be of about 8 $/t P_2O_5 if it is admitted that the whole complexant used is destroyed and of about 5,5 $/t P_2O_5 if it is supposed that the 2/3 of the complexant is regenerated. (As an information, note must be taken that the cost price of 1 t of P_2O_5 is about 300 $).

6 - Conclusions

So we have demonstrated that the ionic flotation technique enables the elimination of the cadmium contained in the wet process phosphoric acid (H_3PO_5 5,5 M ; 28 % P_2O_5). All the results obtained on synthetic and industrial solutions point out the following :

- It is possible to float cadmium with the sodium diethyldithiophosphate (LET) in H_3PO_4 5,5 M and 11,5 M media although this operation is difficult in a diluted aqueous solution. In phosphoric acid 5,5 M, the collector excess is soluble, but we think that an acid filtration on phosphogypsum (Uranium-Pechiney process) could eliminate it by adsorption.
- Although Cu^{2+} ions precipitate but do not float with the LET collector it is possible to float the cadmium in the presence of copper. This latter is quantitatively carried away by the cadmium precipitate. In this present case, we must use a collector excess.
- If we operate at a temperature lower than 25° C, the presence of high concentration ferric ions (10^{-1} M) has a moderate influence on the cadmium flotation efficiency (on the copper one, too). The use of an

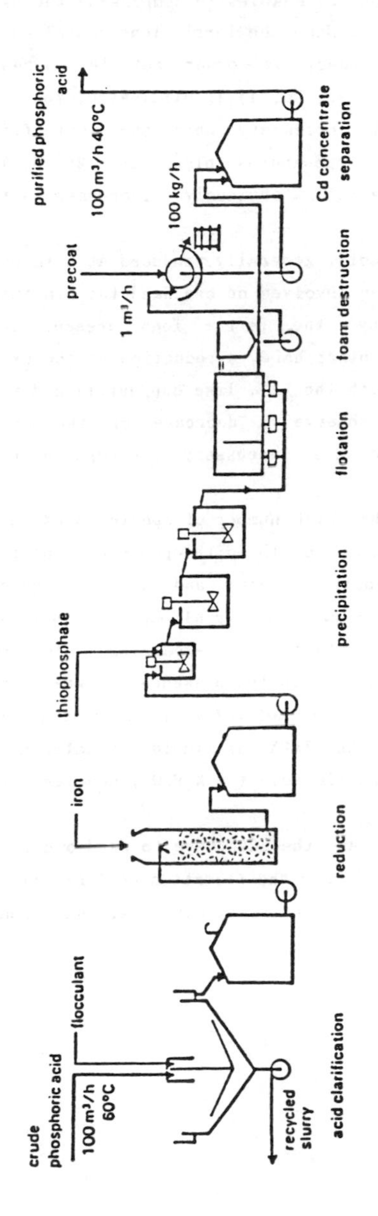

Figure 6 - Flow-sheet of cadmium ionic flotation from 30 % P_2O_5 wet process phosphoric acid.

excess of collector (Φ about 6) enables to compensate the oxidized part by Fe(III). We can also reduce the ferric ions contained in the phosphoric acid with iron powder, in order not to increase the collector consumption. In this case, Cu(II) is eliminated, too.

- The ferric ions reduction is necessary when the ionic flotation operations are realised at temperatures higher than 25° C. The LET collector oxidation by Fe(III) obviously increases with the temperature.
- In the industrial phosphoric acid, generally produced at a temperature of about 60° C, the reduction involves on one hand the limitation of the collector oxidation by the ferric ions present in high concentrations, and on the other hand, a reduction of the oxidizing species susceptible to react with the LET, like copper(II) and vanadium (V). As a result, we can observe a decrease of the collector consumption in comparison to the necessary consumption when no reduction.
- Finally and in spite of the high number of species present in the industrial phosphoric medium, diethyldithiophosphate appears as practically selective in reduction medium and would represent an obvious interest for an eventual industrial application, as it is cheap. We must emphasize too, the fact that the cadmium ionic flotation in TAIBA phosphoric acid is realised in the presence of organic matter, which constitutes an advantage on the solvent extraction technique.

By another way, the use of the IMOX gas-liquid contactor as bubble generator in wet process phosphoric acid (30 % P_2O_5) enables the ionic flotation execution.

Moreover, the IMOX apparatus has the advantage to need a quite small space and can be set anyhwhere below the flotation cell or straight up over the phosphoric acid solution with its outlet situated under the solution's surface.

REFERENCES

1. M.Hutton, Phosphorus and Potassium, 1983, 123, 33
2. V.P. Helmold, B. Kurt and B.H. Tuisko, Chemical Abstract, n° 98 218081 x)
3. K. Frankenfeld, P. Rushke, P. Brodt and G. Eich, European Patent, E.P. 94 630 11 (DE 3 918 599, 18/5/82), 23/11/83
4. V.P. Helmold and S. Guenther, Chemical Abstract n° 99 21485 w
5. S. Guenther, K. Werner and G. Reinhard, Chemical Abstract n° 200 676 k)
6. S. Guenther, G. Reinhard and H. Gero, Chemical Abstract n° 100 105 884 c
7. G. Reinhard, S. Guenther, K. Werner and H. Gero, Chemical Abstract n° 99 142 426 m
8. P. Blazy, E. Jdid, French Patent F 2530 161 (A1) [82 12450], 16/07/82
9. E. Jdid, J. Bessiere and P. Blazy, Revue de l'Industrie Minérale "Les Techniques", 1984, 389
10. M. Lebon, M. Prevost, E. Jdid and P. Blazy, French Patent 84 18301, 30/11/84
11. E. Jdid, P. blazy, M. Lebon, M. Prevost, R. Durand and J. Saujet, XVth International Mineral Processing, June 1985, t.II, 144
12. J. Bessiere, P. Blazy, E. Jdid and A. Floreancig, French Patent 83 21 039, 23/12/83
13. F. Sebba, "Ion flotation", American Elsevier, New York, 1962
14. T.A. Pinfold, *"Ion flotation" in adsorptive bubble separation techniques, R. Lemlich, Academic Press, New York, 1972, chap. 4, p. 53, *"Precipitate flotation", Idem, chap. 5, p. 75
15. A.N. Clarke and D.J. Wilson, "Foam flotation". "Theory and Applications", Marcel Dekker Inc., New York, 1983
16. E. Jdid, J. Bessiere and P. Blazy, Rapport D.G.R.S.T., Nov. 1982, action concertée VRSS n° 80.7.0422 and 80.7.0423
17. J. Bessiere, M. Bruant, E. Jdid and P. Blazy, International Journal of Mineral Processing, 1986, 16, 63
18. E. Jdid, P. Blazy and J. Bessiere, Colloque Sophia Antipolis, 1982, Documents B.R.G.M., 61, 317

19. E. Jdid, P. Blazy, J. Bessiere, C. Tracez and P. Haïcour, French Patent F.2.535.217 (A1) [82.181 142], 29/10/82
20. G.W. Poling, "Reactions between thiol reagents and sulphide minerals", Am. Inst. Min., Metal. & Petrol. Eng., A.M. Gaudin, Memorial Volume, 1976, Vol. 1, p. 334
21. D.W. Fuerstenau and R.K. Mishra, Congrès de Rome, The Inst. of Min. and Metal., 1980, p. 271
22. French Patent F 2.484 862 (80 13527), 18/06/80

Metal Removal Using Coordinating Copolymers

By M.J. Hudson

CHEMISTRY DEPARTMENT, UNIVERSITY OF READING, WHITEKNIGHTS, PO BOX 224, READING, BERKSHIRE, RG6 2AD, UK

Abstract

Coordinating copolymers are used in a wide variety of processes for the recovery of metals from aqueous and non-aqueous solution. The principal areas of use are in the recovery of toxic and precious metals particularly from dilute solution. They may also be used for the recovery of radionuclides from liquid nuclear effluent.

Introduction

Coordinating copolymers are also known as selective ion-exchangers. They are polymers in which ligands are covalently bound to a central chain or matrix. Such materials may be of natural or synthetic origin, inorganic or organic, soluble or insoluble. The term coordinating copolymers is frequently used for synthetic, insoluble, organic-based macromolecules and the discussion which follows will principally, but not exclusively, concentrate on this type of compound. In addition to the functional group (active site) on the polymer, it is necessary to consider the nature of the polymer chain (e.g. hydrophobicity, degree of crosslinking). The chemistry of the metal in the particular aqueous environment also is important.

Some Aspects of the Fundamental Chemistry

The application of coordinating copolymers to the separation and extraction of metals is not new as one of the first patents was filed in 1947. This was for a dipicrylamine resin[1] for the extraction of potassium. Some chemical and chemical engineering aspects of these materials have been reviewed.[2-8] The principal concerns with respect to the industrial application of these macromolecules are of course connected with the economics. Their cost is higher than traditional ion-exchangers and sometimes the kinetics of extraction and elution are slower. It appears that coordinating copolymers may, therefore, only be used when other reagents have been demonstrated to be unsuccessful. The areas of application include removal of metals (ions) from strong acid solution in which traditional ion-exchangers are ineffective, recovery of toxic metals, recovery of precious metals and extraction of radionuclides. In addition, it has to be said that there are considerable developments in the use of coordinating copolymers in areas far removed from the field of metal extraction processes. These include adhesives[9], catalyst supports and modifiers for the surfaces of electroactive materials. Each of these applications employs the high affinity of metal ions for the active sites on the coordinating copolymer.

Some of the available commercial resins are listed in Table 1. The ten principal types of commercial resins have been tested but the range of reported laboratory reagents is well over one thousand. Each resin manufacturer usually has an aminodiacetate resin and more are developing aminophosphonic acid and polyamine resins. The donor atoms refer to those which may covalently bind

TABLE 1

Some of the Principal Coordinating Copolymers.

Functional Group	Nature of the Group	Donor Atoms	Trade Name	Company
1. $CH_2.N(CH_2CO_2H)_2$	Iminodiacetate	$(N)O_2$	ES 466 Dowex A-1 Chelex-100 IRC-718 Wofatit MC50 TP 207	Duolite Dow Biorad Rohm and Haas VEB Bayer
2. $CH_2NCH_2PO(OH)_2$	Aminophosphonic Acid	$(N)O_2$	ES 467	Duolite
3. $CH_2N.(CH_2.CHOH.CH_3).CH_2C_5H_5N$	Weak Base	N_2	XFS 43084	Dow
4. $-CH_2N.\overset{R}{C}H.C_5H_5N$	Weak Base	N_2	XFS 4195	Dow
5. $-CH_2N(CH_2CH_2N)_n-H$	Polyamine	N	CR 20	Mitsubishi
6. $-C_5H_5N$	Weak Base	N	CR 2	Sumitomo; Reilly SnChem
7. $-CH_2.SCNH.NH_2$	Isothiouronium	N_2	Srafion NMRR Ionac SR-3	Ayalon Ionac
8. -SH	Thiol	S	IMAC TMR	Sybron
9. $-NCS_2H$	Dithiocarbamic Acid	NS_2	Misso ALM 525 Q-10R	Nippon Soda Sumi Chemicals
10. Cryptand 221B	Cryptand	NO	Kryptofix 221B	Parish

to the metal. Some such as the thiols are monodentate (one donor atom per ligand); some may be bidentate e.g. N_2 for XFS 4195 or polydentate like the polyamines. Superimposed over number of dentate groups per ligand is another called the 'polymer effect' which arises especially in low crosslinked resins because each polymer chain acts as a polydentate ligand. Thus the elution of metals may be difficult as free donor sites can bind to the metal. Elution is possible using higher acid concentrations, polymer of higher molecular mass or a metal higher in the Irving-Williams series.

The stereochemical requirements of metals (e.g. tetrahedral coordination) must be satisfied. Consequently, the polymer chain and the pendant groups should be flexible. This is ahieved by having little or no crosslinking, flexible e.g. long (alkyl) chains and pendant linkages. The capacity of the coordinating copolymer can be maximised by having the greatest number of available active functional groups. The word 'available' has been used advisedly because the metal ion must have good access to the donor groups. This is achieved by synthesis of the macromolecule and by using a small bead size. Coordinating copolymers may be used with advantage as powders as these have a maximum surface area. The covalent bonds formed between the metal and copolymer may act as crosslinks and prevent diffusion of the metal ion. Thus it has been shown that the rate of extraction of rhodium(III) by a dithiocarbamate resin[10] is greater with small particles than large ones. The time for half the extraction ($t_{\frac{1}{2}}$) was 15 mins for particles over 500 micron and six minutes for 150-250 micron size range. The capacity of the smaller particles was also greater (0.6 mmoles g^{-1}

compared with 0.29 mmoles g^{-1} of dry polymer.) The loaded polymer was readily filterable as the rhodium crosslinked the copolymer and gave a heavy and dense material.

Short Bed Ion-Exchange Technology

Coordinating resins in the powdered form may be used in short bed ion-exchange equipment. The Recoflo ion-exchange system[11] is commercially available for such resins and this uses resin columns 5-60 cm long which is, of course, much shorter than the normal ion-exchange column. The resin inventory is low and liquid phases are pumped through the resin. There is counter current regeneration and short cycle times. These and other features enable the principal advantages of the coordinating copolymers to be exploited. A flow-sheet for the separation of

FIGURE 1

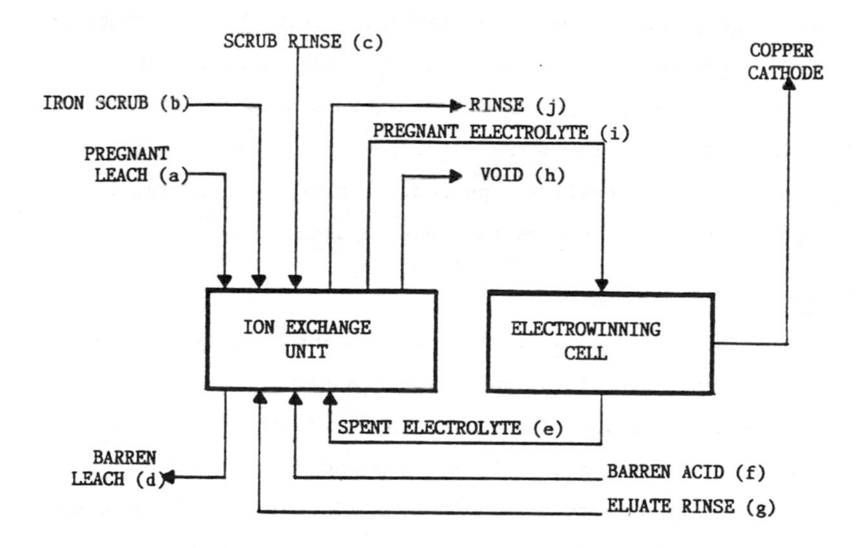

Process Flow-Sheet for the Separation of Iron(III) and Copper(II)

copper(II) from iron(III) is shown in Figure 1. Copper(II) has an affinity for amine or nitrogen containing resins. The active site and the stereochemistry of nitrogen atoms around the copper can be determined as indicated later. Weak based resins such as XFS 43084, XFS 4195[11,12] CR 20 or CR 2 are suitable for such separations. In Figure 1, the pregnant solution is pumped through the resin and copper with some of the iron is extracted(1). Iron is scrubbed with a dilute solution of sulphuric acid(2). Water is used to rinse off the residual acid and iron and spent electrolyte elutes copper. Barren (copper-free) sulphuric acid leaches residual copper which can then pass to an electrolytic cell. A pregnant feed of Cu-Fe in the ratio 1:2 can be thus processed to give an electrolyte of composition (Cu:Fe;20:1). The resin inventory was over twenty times lower than a continuous counter current ion-exchange process. The processes using short bed ion-exchange technology are capable of extension to a wide range of other separations. The resin inventory is minimised as the volume of resin used is small and the attrition is minimal because liquids and not the resin are pumped. Resins are reported to be capable of reuse for more than one thousand cycles.

Industrial Applications

One of the principal uses of chelating ion-exchangers is connected with the removal of calcium and magnesium from the brine feed for chlor-alkali cells. The aminophosphonic acid resins, such as Duolite 467, are capable of reducing the calcium and magnesium to a few parts per billion as required for the membrane cells.[14,15] The iminodiacetic acid resin is used to separate nickel from a cobalt(III) pentammine solution. Nickel

(800 mg dm^{-3}) can be removed from a solution of the cobalt(III) complex.[16] The coordinating resins can be highly selective for the removal of traces of one metal from higher concentrations of another. Thus traces of copper can be removed from cobalt electrolytes.[17-22] Traces of nickel and copper may be removed using the picolylamine (2-aminomethylpyridine) resin (XFS 4195) at Inco's Port Colborne refinery in Canada.[11,12] A similar resin[20] has been used to recover copper from iron-containing acid leach liquors. The high affinity of copper for amines is exploited to separate copper(II) from a solution containing Fe(II) and Fe(III). The study has been carried out on a pilot scale and 300 dm^3 h^{-1} were treated to give 250 g h^{-1} of copper (weighed at a cathode). Nickel[23] can be removed from process effluents. Zinc[24] may similarly be extracted.

A comparative study of some chelating resins has been made[25] and it was concluded that the principal use of chelating ion-exchangers in hydrometallurgy will be in the recovery of metal values from dilute solution. At higher (say 200 ppm) concentrations, solvent extraction often has a clear economic advantage unless resin regeneration is fast. With respect to the separation of copper from iron in acid sulphate media, the picolylamine resins and the aminophosphonic acids show a moderate selectivity for copper over iron. However, the capacity of the aminophosphonic acid appears to be low so that the picolylamine resin is favoured.

Active Site Determination

The active sites to which metals bind may be determined in favourable cases using ultraviolet and visible spectroscopy or electron spin resonance. The selective ion-exchanger acts as a polydentate ligand. For a given set of conditions such as pH or metal ion concentration it appears that the metal adopts one (or two) stereochemistries within the polymer.[26] This stereochemistry may change with loading. For example with an N-dithiocarbamate polymer (NCS_2H) the preferred site is five coordinate at 1% loading but at higher loadings appears to be CuS_2N_2 and is a model for copper 'blue' proteins. There has been considerable effort to establish the composition and geometry of copper binding in polymeric systems. The electron spin resonance spectra (esr) have been shown to be diagnostic.[27] in the X-band spectrum the A-parallel and g-parallel (Figure 2) values can be related to the donor atoms in the coordination

FIGURE 2

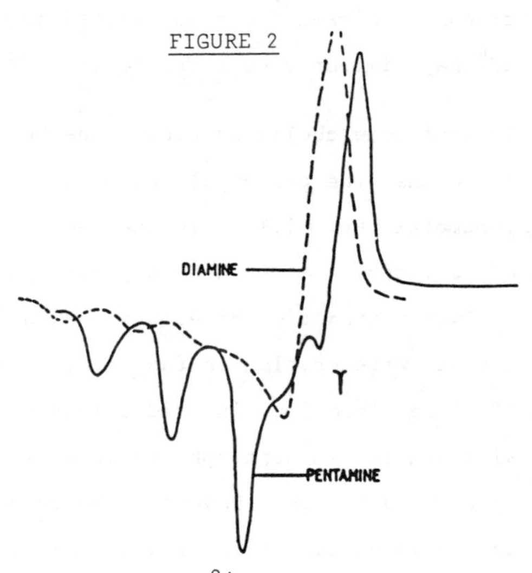

Electron Spin Resonance of Cu^{2+} - amine polymers. The amine groups are covalently bound to the polymer.

sphere. Hudson and Matejka[28] have studied the esr spectra of a series of polyamine groups which are covalently bound to the styrene or acrylic backbone. The results are listed in Table 2.

TABLE 2

Stereochemistry of Copper in Some Polymers with Polyamine* side Chains.

Functional Group	n	$A_{//}$	$g_{//}$	Inference of Stereochemistry
1. Acrylic Base Polymers				
$-CONH(CH_2CH_2NH)_n.CH_2CH_2NH_2$	0	175	2.26	N_2O_2; ca 30°
	4	170	2.23	N_4; $20-30^o$
	5	170	2.23	N_4; $20-30^o$
2. Styrene Based Polymers (gel - 2% DVB)				
$-NH.(CH_2CH_2NH)_n.CH_2CH_2NH_2$	1	220	2.17	N_4; perhaps N_2O_2 or N_3O square planar (s.p)
	2	183	2.21	N_4; s.p.
	3	173	2.21	N_4; s.p.
	4	170	2.20	N_4; s.p.

*polyamine is used in the sense that there are more than one N atom in the side chain.

In the case where there is a short side chain, e.g. $CONHCH_2CH_2NH_2$, there is strong evidence from the g-parallel value that one or more oxygen atoms are bound to the copper. However, when there are more nitrogen atoms in the side chain the copper appears to be only bound to nitrogen. It appears likely that in both the acrylic and styrene polymers more than one side chain is required to bind to the copper. It is not clear at this stage whether the binding is intra- or inter-polymer strand. The values of the parameters are remarkably constant for n greater than three. Consequently, it is likely that n should be three or more for a commercial selective ion-exchanger. The angles in the last column refer to the dihedral angle through which two of the donor atoms have been twisted from the square planar positions. A typical spectrum is shown in Figure 2.

Formation constants, the number of active sites and mole fraction of those sites occupied can be derived from titration curves of the metal against the polymer. One technique used for plotting the results is the Scatchard method which uses the relationship

$$\frac{[\bar{n}]}{[L]} = n\,K - \bar{n}\,K$$

$[\bar{n}]$ = mole fraction of site occupied

$[L]$ = activity of the free metal in solution

K = conditional formation constant

n = number of sites.

Thus there is one straight line of gradient - K for each site when $[\bar{n}]/[L]$ is plotted against $\bar{n}$. This technique has been used[29,30,31] for mixed amine and dithiocarbamate polymers and it has been shown that there were two well defined sites on the polymer for which the idealised formula is given below. The formation constants and (mole fraction) of 10^{11} (0.14) and

$$\left[\begin{array}{l} CH_2CH_2\overset{+}{N}H.CH_2CH_2NH \\ \qquad\qquad | \\ \qquad\quad CH_2CH_2NHCS_2^{-} \end{array}\right]_n$$

10^5 (0.22) were related to binding to $-NCS_2H$ and amine groups respectively. The method can be extended to a wide range of polymeric materials and inorganic as well as organic backbones can be included. Inorganic reagents appear to have faster kinetics than organic based ones as the sites are on the surface of the inorganic matrix. A typical functionalised silica[32] is shown in Figure 3 and the Scatchard plot (Figure 4) shows

FIGURE 3

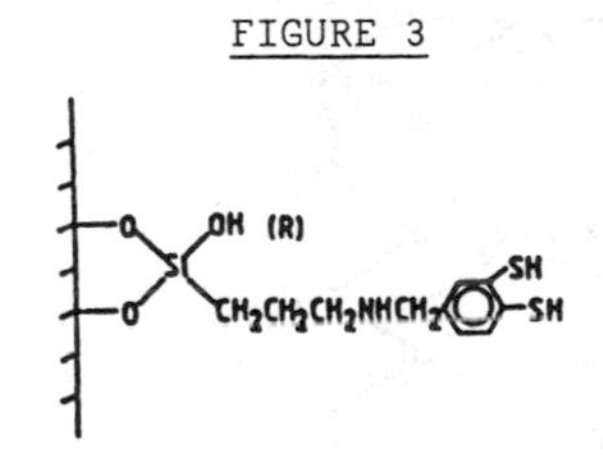

Functional group bound to silica gel by a silylation reaction (after Kettrup).

FIGURE 4

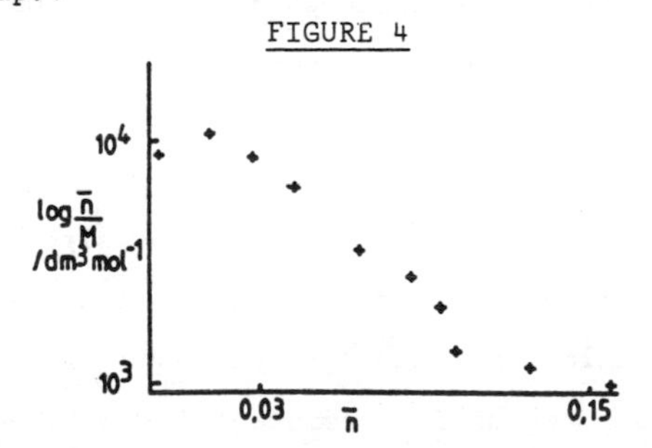

Scatchard plot; copper ion selective electrode; nitrate media.

FIGURE 5

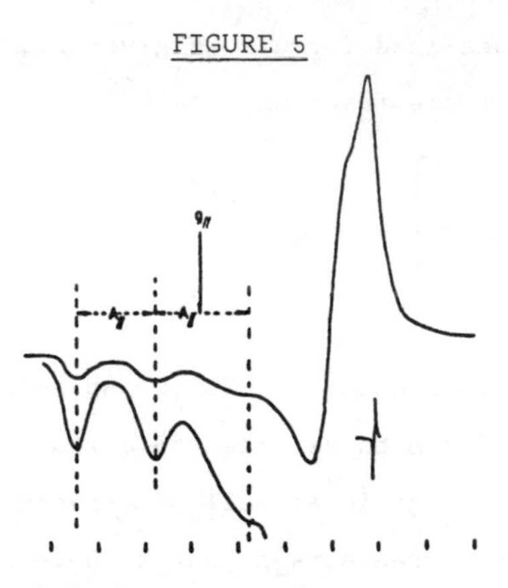

X-band Electron Spin Resonance for copper on the functionalised silica.

FIGURE 6

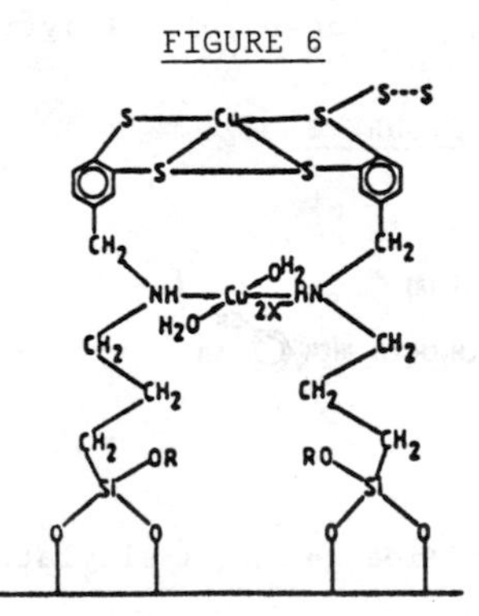

Possible stereochemistries for the copper. (N_2O_2 coordination implies two chains binding to one copper).

---

two clearly identifiable sites. The ratio of the equilibrium constants is 31 and the ratio mole fractions of sites occupied is 2.26 (both amine: sulphur). These two sites are associated with amine and thiol groups respectively. The esr spectrum (Figure 5), which is consistent for a range of loadings, indicates N_2O_2 coordination of a copper(II) species. The possible sulphur site which is depicted in Figure 6 is not seen in the esr as this is a copper(I) (d^{10}) site. It appears from these pieces of evidence that there is significant binding to the nitrogen atoms. Further studies are in progress as the nature of the binding is important in view of the interest of the nuclear industry in this type of material.

Nuclear Effluent Treatment. Low activity liquid nuclear waste effluent may be treated using coordinating copolymers. Thus ^{106}Ru may be extracted using a range of sulphur or nitrogen based compounds. A thiol resin[33] has a high K_D value (1550 from a solution with Ru-NO species 3.63×10^{-6} M; HNO_3 0.15 M; $NaNO_3$ 5 M) for the extraction of ^{106}Ru from an acid solution with a high salt concentration. Under such conditions ion-exchange resins and zeolites are ineffective. Interestingly, from the same solution composition the iminodiacetate resin (K_D 800) was also effective. The chelate effect seems to favour proton loss as is the case with copper in solvent extraction systems. With simulated pond water (NaOH 0.005 M; Na_2CO_3 0.0028 M; $NaNO_3$ 0.01 M; Ru-NO species 3.63×10^{-6} M) the dithiocarbamate polymers have a very high K_D value (13800) so that this type of material could be used in treating nuclear wastes. Further work in this area is in progress.

Toxic Metals. These include cadmium and mercury. Both can be extracted by normal ion-exchange resins when they are present as simple M^{2+} species. In environmental waters they are frequently present in colloidal species which can effectively be extracted using hydrophobic interactions on a styrene adsorber. Soluble complex species of these metals can be immobilised on coordinating copolymers to polish the solutions after liming or floc treatment[29-34]. The polymer may be used as a gel and the metal-loaded xanthate precipitates out. These reagents are inexpensive to produce and mobile plant can go to the required treatment site. The dithiocarbamate polymer[29] has a higher capacity but there is concern about the toxic nature of the poly(etheneimine) itself. Cadmium can be removed from phosphoric acid (3 M) media[35] using coordinating resins. Trace quantities of mercury can be extracted using thiol resins.

Precious and Platinum Group Metals. Gold(III) has an affinity for oxygen ligands such as polyethers like poly(ethene oxide). In the strong acid media used in refining (5 M HCl) the ion-exchange mechanism dominates. The coordinative mechanism occurs at higher pHs which are often avoided to limit the precipitation of the metals and the irreversible binding to coordinating resins. Much of the fundamental aspects of the application of selective ion-exchangers to the extraction of precious and platinum group metals has been recently reviewed.[6,7,8] In addition the application of poly(4-vinylpyridine) to the separation of rhodium and iridium has been studied.[10] The copolymer was cross-linked with 6% DVB and was effective[36] at separating iridium(IV) from rhodium(III) in hydrochloric acid. The metal concentration in the effluent expressed as a percentage of output/input is given in Figure 7. The iridium is extracted as $[IrCl_6]^{2-}$ and the $[RhCl_6]^{3-}$ chloroanion is not extracted. Elution of the iridium is difficult and if left to stand the pyridine coordinates to the iridium(IV) and there seems to be a reduction to iridium(III).

FIGURE 7

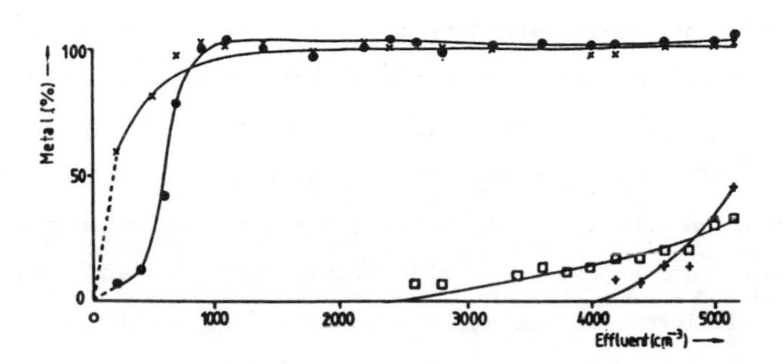

Separation of Rh •; Cu X; Ir □; Pt +
From palladium raffinate solution with poly(4-vinylpyrdine) using 4 M hydrochloric acid.

There has been an increase in interest in using selective ion-exchangers for scavenging precious and platinum group metals. Among the many that have been tried, the following are of industrial potential. N-vinylalkylpyrazole[37,38]; S-dithizone[39]; aminoquinoline[40]; 8-mercaptoquinoline and rhodamine[41,42]; sulphophenylmethylpyrazolones[41,42]; aminoazo[43]; thiocyanate[44]; amide[45]; amidoxime[46]; thioglycolate[47]; sulphonylguanidine[48]; glyoxal-bis-(2-mercaptoanil)[49]; 2-mercaptobenzimidazole[49]; hydroxamic acid[50]. The styrene, silica and acrylic matrices were used but preference seems to be for the poly(styrene)-based polymers as these are chemically stable under the conditions of use.

Future Industrial Use

The industrial use of selective ion-exchangers may rely wholly on one unit of equipment such as the Recoflo system (Figure 1)[51] or a continuous ion-exchange column.[52] However, it is possible that they could be used as a unit (or part) of a larger flow-sheet. Thus selective ion-exchangers could be used for polishing raffinates from solvent extraction; improving industrial effluent which contains toxic metals; treating environmental waters; extraction processes using dilute solutions and treating plating solutions. In the nuclear industry they could be used after floc treatments, in or after ultrafiltration processes. However, as stated earlier, coordination polymers have found wide application in areas other than extraction processes. These applications include adhesives[9,53], oil additives and surface modifiers.[54] It is anticipated that there will be an increased market for all applications.

References

[1]A. Skogseid, Norwegian Patent 72, 583 (1947).

[2]J.R. Millar, Chem. Ind., 1957, 606.

[3]M. Kaneko and E. Tsuchida, Macromol. Rev., 1981, 16, 397.

[4]C. Calmon, Reactive Polymers, 1982, 1, 3.

[5]K. Sahni and J. Reedjik, Coord. Chem. Revs., 1984, 59, 1.

[6]M.J. Hudson, "Coordination Chemistry of Selective Ion-Exchange: Science and Technology", NATO ASI Series E. Applied Sciences, 107, Martinus Nijhoff, Dordrecht, 1986, 35.

[7]ibid., "Some of the Uses of Selective Ion-Exchangers in Hydrometallurgy", 463.

[8]A. Warshawsky, ibid., 67.

[9]L.H. Lee (ed.), "Adhesive Chemistry Developments and Trends", in "Polymer Science and Technology", New York, Plenum, 1984. 29.

[10]A.F. Ellis, Ph.D. Thesis, Reading (1985) - (some restrictions on loans).

[11]R.R. Grinstead, J. of Metals, 1979, 31, 13 (see also ref. 51).

[12]R.R. Grinstead, International Conference "Ion-Exchange Technology", Cambridge IEX 84, Ellis Horwood, 1984, 509.

[13]K.C. Jones and R.A. Pyper, J. of Metals, 1979, 31, 19.

[14]J.J. Wolff, "Ion-Exchange Purification of Reed Brine for Chlor-alkali Electrolysis Cells, The Role of Duolite ES 467", Oslo Symposium, "Ion-Exchange and Solvent Extraction", London Soc. Chem. Ind., IV/62-IV/74, 1982, Section IV, 62.

[15]O. Langeland, ibid., IV/52-IV/61

[16]C. Nikolic, J.L. Blanco, R. Crnojevich and W.E. Sherwood, "The AMAX Port Nickel Refinery Process for Cobalt Purification and Recovery", Process Fund. Considerations Sel. Hydromet. Syst., M.C. Kunn ed., Soc. Min. Eng. AIME, 1981, 115.

[17]B. Strong and R.P. Henry, The Purification of Cobalt Advance Electrolyte using Ion-Exchange, Hydrometallurgy, 1976, 1, 311.

[18]B.R. Green and R.D. Hancock, Useful Resins for the Selective Extraction of Copper, Nickel and Cobalt, J.S. Afr. Inst. Min. Metall., 1982, 82, 303.

[19]F.A. Vernon, Some Aspects of Ion-Exchange in Copper Hydrometallurgy, Hydromet., 1979, 4, 147.

[20]R.R. Grinstead, W.A. Nasutavicus and R.M. Wheaton, New Selective Ion-Exchange Resins for Copper and Nickel, in "Extractive Metallurgy of Copper", eds. J.C. Yannopoulos and J.C. Agarwal, New York: AIME, 1976, 2, 1009.

[21]R.R. Grinstead and K.C. Jones, Properties and Hydrometallurgical Applications of two New Chelating Ion-Exchange Resins, Chem. Ind., 1977, 637.

[22]D. Naden and G. Willey, "Reduction of Copper Recovery Costs Using Solid Ion-Exchange" in "Theory and Practice of Ion-Exchange", Society of Chemical Industry, London, M. Streat, ed., 1976, 44.1–44.12.

[23]H.W. Kauczor, Recovery of Nickel by Ion-Exchange, I. Chem. Ed. Symposium Series, No.42, London: Inst. Chem. Engineers, 1975, 20.1-20.3.

[24]H. Vejima, M. Hirai and T. Ishibashi, Recovery of Zinc from Industrial Waste Waters using Chelating Resins, Prog. Water Tech., 1977, 9, 871.

[25]J. Melling and D.W. West, A Comparative Study of Some Chelating Ion-Exchange Resins for Applications in Hydrometallurgy, Ion-Exchange Technology, D. Naden and M. Streat (eds.), Soc. Chem. Ind., London, 1984, 360.

[26]A.F. Ellis, M.J. Hudson and A.A.G. Tomlinson, J. Chem. Soc., Dalton Trans., 1985, 1655.

[27]H. Tokoi and A.W. Addison, Inorganic Chemistry, 1977, 16, 1341.

[28]M.J. Hudson and Z. Matejka - previously unpublished results.

[29]A.F. Ellis, M.J. Hudson and G. Tiravanti, Makromol. Chem., 1985, 186, 339.

[30]R.F.C. Mantoura and J.P. Riley, Anal. Chim. Acta, 1978, 78, 193.

[31]G. Sposito, Environ. Sci. Technol., 1981, 15, 396.

[32]M.J. Hudson and A. Kettrup submitted for publication, 1986.

[33]A. Dyer, D. Keir, M.J. Hudson and B.K.O. Leung, J. Chem. Soc. Chem. Commun., 1984, 1457.

[34]D. Marani, M. Mezzina, R. Passino and G. Tiravanti, Environ. Tech. Letters, 1980, 1, 141.

[35]M.J. Hudson patent applied for.

[36]J. Barnes, A.F. Ellis and M.J. Hudson, Angewandte Chemie (submitted for publication).

[37]G.V. Myasoedova, Talanta, 1976, 23, 866.

[38]A.L. Braslavskaya, U.S. Patent 4,093,792, 1978.

[39]M. Grote and A. Kettrup, "Ion-Exchange Technology", Conference Cambridge, D. Naden and M. Streat Ed., Ellis Horwood, 1984.

[40]S.B. Savvin, I.I. Antokoskaja, G.V. Myasoedova, L.I. Bol'shakova and O.P. Shvoeva, J. Chromat., 1974, 102, 287.

[41]G.V. Myasoedova, L.I. Bol'shakova, O.P. Shvoeva and S.B. Savvin, Z. Anal. Khim., 1971, 26, 2081.

[42]G.V. Myasoedova, L.I. Bol'shakova, O.P. Shvoeva and S.B. Savvin, Z. Anal. Khim., 1972, 27, 2004.

[43]G.V. Myasoedova, L.I. Bol'shakova, O.P. Shvoeva and S.B. Savvin, Z. Anal. Khim., 1973, 28, 1550.

[44]G.A. Kanert and C. Chow, Anal. Chim. Acta, 1975, 78, 375.

[45]C. Phlandt and J.S. Fritz, J. Chromat., 1979, 176, 189.

[46]Diamond Shamrock Chem. Co., Duolite CS-346 Tech. Sheet, 1972.

[47]J.S. Fritz, Pure Appl. Chem. 1977, 49, 1547.

[48]A. Gulko, H. Geigenbaum and G. Schmuckler, Anal. Chim. Acta, 1972, 59, 397.

[49]E. Bayer, Angew. Chem. Int. Ed., 1964, 3, 325.

[50]F. Vernon and M.Z. Wan, Anal. Chim. Acta, 1981, 123, 309.

[51]C.J. Brown, D. Davy and P.J. Simmons, Plating and Surface Finishing, 1979, 66, 54.

[52]M. Streat (see reference 6).

[53]P.M. Hergenrother, "High-temperature Adhesives", Chemtech, 1984, 496.

[54]E. Killman, Angew. Chem. Int. Ed. (Eng.), 1974, 13(6), 415.

The Significance of Mercury in Combustion Processes

By D.O. Reimann

GARBAGE INCINERATION AND POWER-HEATING PLANT, CITY AND COUNTY OF BAMBERG, RHEINSTRASSE 6, D-8600 BAMBERG, FRG

1. General Information on Mercury

Besides other heavy metals, e.g. cadmium, lead, nickel, varying quantities of mercury and its compounds occur in the combustion product during garbage incineration. Different mercury compounds can be formed and decomposed again due to the continuously changing composition of the flue gas as well as the increasing and decreasing temperatures during the combustion process.

Mercury, a liquid metal in element form at room temperature, occurs as well in alloys and in inorganic and organic compounds. A precise description of the solubility characteristics, chemical reactions and sequences for Hg is not possible within the scope of this report. Detailed and extensive information is given in the literature (e.g. 1 - 6). However it does appear to be practical to quickly describe the detection and analysis methods.

Mercury and its compounds must be detected in solid, liquid and gaseous form. Aqua regia detection can be used for solid samples. Closed analysis with sulphuric or nitric acid and potassium peroxodisulphate ($K_2S_2O_8$) has proved very effective for determining the total content of mercury in liquid samples and from the gas phase. After analysis the mercury is present in a dissolved state in bivalent compounds.

Quantitative analysis of the mercury is accomplished according to the cold vapor technique described in DIN 38406 using atom absorption spectrometry (AAS).

Non-representative sampling, difficulties and expenditures in the detection procedures and the high volatility of Hg etc. can lead to erroneous quantitative and qualitative results. Certainly it will require even more extensive and detailed research and analysis in order to go beyond the total quantity of Hg and also provide information on the individual Hg compounds in a sample. This applies particularly to the Hg distribution during the incineration of the heterogeneous product garbage.

Information on organic Hg compounds is not given within the scope of this paper, because organic Hg compounds can be expected only in extremely low concentrations in garbage incineration - if at all. The compounds are thermically unstable and decompose generally at 300 - 400°C.

2. Establishment of Mercury Limits for the Environment

In the Federal Republic of Germany the laws, regulations, pollution codes and restrictions regarding mercury and its compounds in waste water and sewage sludge as well as workplace pollution and emission prescribe extremely low limits.

2.1 Mercury in Waste Water

The Community Waste Water Contamination Codes based on the presently valid regulatory work A 115 of the Technical Waste Water Union (Regelwerk der Abwassertechnischen Vereinigung (AVT), 1983) do not allow concentrations higher than 0.05 mg l^{-1} for mercury. Cadmium is classified on the second place for problematic heavy metals with 0.5 mg l^{-1} and the permissible concentrations are therefore 10 times as high.

The Federal German Waste Water Tax Code (10) also emphasizes the danger of Hg to such an extent that under maintenance of the permissible pollution values introduction or disposal per kg Hg into a recipient is calculated at 50 pollution units - equal to DM 2.000 per kg Hg. By comparison the market price for 1 kg Hg is between DM 150 and DM 400, depending upon the purity. In the second place the dangerous heavy metal cadmium is listed with 10 pollution units per kg.

2.2. Mercury in Sewage Sludge

Heavy metal ions are partly adsorbed at organic substance of the sludge and therefore can lead to problems when the sludge is intended for agricultural use. According to the present valid German Sewage Sludge Regulation a max. concentration of 25 mg Hg per kg dry mass is allowed in sludge from sewage plants.

2.3 Mercury Pollution at Workplaces

Mercury can be absorbed by the human organism by inhalation of vapors and dusts, by the stomach-gut-tract and through the skin. In order to avoid the possible danger presented for humans by mercury, limits are established for mercury in the Federal Republic of Germany as follows:

Upper standard limit: Definition according to (12) as a value, which is not exceeded by 95 % of the average population; 5 µg Hg l^{-1} in blood or 5 µg Hg g^{-1} creatinine in urine (13).

Biological material tolerance value (BMT-value): Established according to (12); up to this tolerance no negative influences to health are to be expected even for exposures during an eight hours day or 40 hours week; 50 µg Hg l^{-1} in blood and 200 µg Hg l^{-1} in urine (13).

Maximum workplace concentration (MWC): 0.1 mg Hg m^{-3} (14)

The following relationship exists between these limits on the basis of extensive tests for Hg:

MWC_{Air}	BMT_{Urine}	BMT_{Blood}
100 μg m^{-3}	200 μg l^{-1}	50 μg l^{-1}
1	1	0.5

"Under critical eluation of the previous findings it is presently assumed that at average Hg concentrations in the air of >0.1 mg m^{-3} health disorders would be possible. At concentrations of <0.1 mg m^{-3} no objective damage resulting from Hg has been reported, with the exception of subjective complaints as well as biological changes without any proven disease value. In comparison with the limits discussed on the basis of dosage effect relationships, these concentrations provide a sufficient safety range up to manifest mercury poisoning" (14).

2.4 Mercury Emissions

The draft for the Technical Codes for Air Pollution 2/86 as an executive order to the Federal Immission Protection Law (15) classifies the heavy metals mercury, cadmium and thallium in the material class I as particularly dangerous. The expected limitation of the permissible emissions in dust form for these 3 heavy metals is 0.2 mg Nm^{-3} at 11 % O_2. However the further stipulations of these codes in paragraph 3.1.4 reportedly state that special precaution of preventing the emissions are required when significant percentages of these materials are present in vapor or gaseous form.

These regulations apply particularly for Hg. In the Federal Republic of Germany a maximum permissible concentration of 0.2 mg Nm^{-3} is be expected in the future for heavy metals, class I, in gas and dust form.

In consideration of MWC value of 0.1 mg Nm^{-3} limitation of the emission value to 0.1 or 0.2 mg Hg Nm^{-3} at 11 % O_2 appears justified. Since, during garbage incineration, emissions become immissions with a dilution factor of $1:10^5$ to $1:10^6$, the immissions are far below the MWC-values, that a threat to health resulting from Hg can be excluded according to the present knowledge in spite of possible concentration in the soil and in the nutrient cycle. In my opinion the very strong appearing mercury limitation is fundamental:

If the mercury emissions in the form of gas and dust, which are the most difficult to detect within the scope of flue gas purification, are bound or reduced, this applies by analogy for all the other heavy metals specified in the technical directions for air pollution, because these are easier to remove than mercury.

Mercury could become an indicator for the reduction of all heavy metals.

3. Garbage Incineration Plant and Associated Mercury Pollutants

Figure 1: A garbage incineration plant with flue gas wet scrubber and residual product quantities. In addition to the heated flue gases with dust concentrations resulting from the prescribed combustion temperature of at least 800°C in the afterburner zone, pollutants and heavy metals released by the combustion are also present.

Flue gas purification therefore has 3 primary objectives:

- Utilization of the free energy in the flue gas
- Collection of the dust particles
- Elimination of the pollutants and heavy metals

For the gaseous heavy metals and compounds, particularly for Hg, the flue gas purification must be accomplished to obtain the best possible condensation conditions via the greatest possible temperature reduction in the flue gas (saturation temperature approx. 60 - 70°C). Frequent Hg reaction products in the flue gas are oxides, chlorides and fluorides, for example. The reactions of the mercury are dependent primarily upon the temperature and pressure as well as on the reduction and oxidation conditions.

3.1 Specific Mercury Distribution for Garbage Incineration

A thermic treatment such as garbage incineration offers a method for determination of the mercury quantity in the garbage. The residual materials such as sludges, filter dusts, waste water from flue gas purification and sludge washing, sorption residuals as well as the pure gases can be collected and analyzed for Hg as relatively homogeneous products much more easily than the initial product, garbage. The pollutant distribution shown in figure 3 for mercury in garbage incineration with flue gas washer is based on the precisely collected quantity data for individual pollutants over a longer period of time, figure 2, and extensive measurements of the Hg concentrations, table 1.

3.2 Effect of Digested Sludge upon the Hg Balance

If both garbage and digested sludge are incinerated together, no problems occur with an addition of up to 10 - 15 % of dewatered digested sludge per ton of garbage, whereby 100 - 150 kg of dehydrated sludge contains approximately 35 kg of dry matter (dm). Incineration of sludge should be accomplished only when the pollutants do not exceed the permissible values according to the sewage sludge codes - for Hg 25 mg $(kg\ dm)^{-1}$ - since the agricultural using of the digested sludge should always be preferred to incineration. If garbage and digested sludge are incinerated together the Hg contents in the incineration product garbage + sewage sludge increases by approx. 1 g mercury per 100 kg of dewatered sludge (17).

25 g of Hg and higher quantities may be present in the flue gas per ton of sewage sludge dry matter, which must be bound during the flue gas purification.

This result indicates the significance of Hg in emissions for incineration of sludge alone.

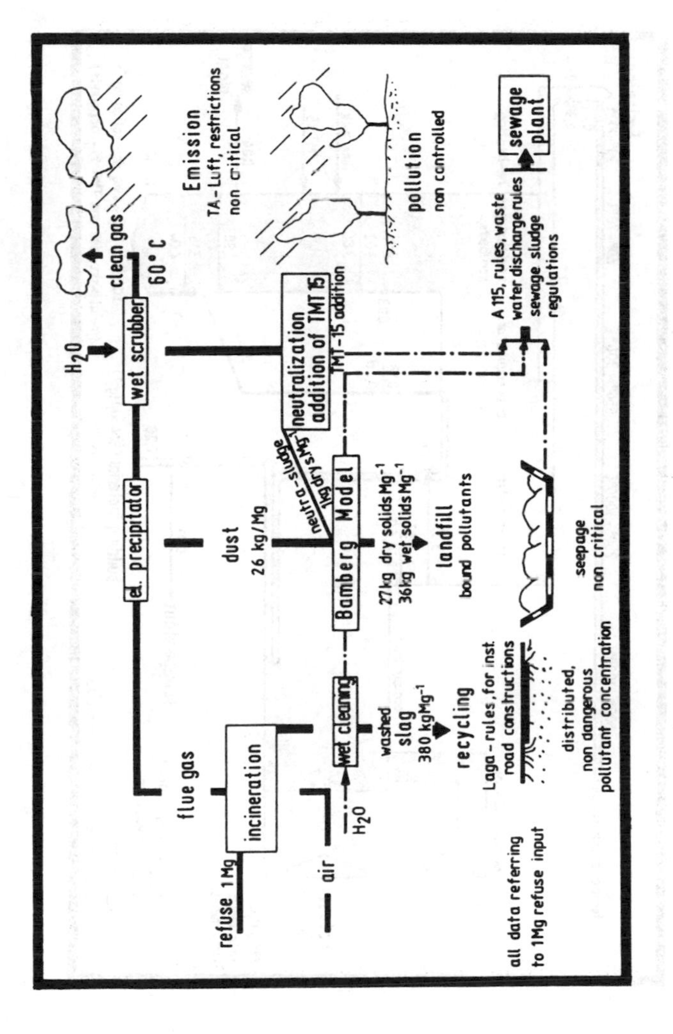
refuse 1 Mg
incineration
flue gas
air
H2O
wet cleaning
washed
slag
380 kgMg-1
recycling
Laga-rules, for inst.
road constructions
distributed,
non dangerous
pollutant concentration
el. precipitator
dust
26 kg/Mg
Bamberg Model
neutra-sludge
1kg drys./Mg
27kg dry solids Mg-1
36kg wet solids Mg-1
landfill
bound pollutants
seepage
non critical
H2O
wet scrubber
clean gas
60° C
neutralization
addition of TMT 15
TMT-15 addition
Emission
TA-Luft, restrictions
non critical
pollution
non controlled
A 115, rules, waste
water discharge rules
sewage sludge
regulations
sewage
plant
all data referring
to 1Mg refuse input

Figure 1

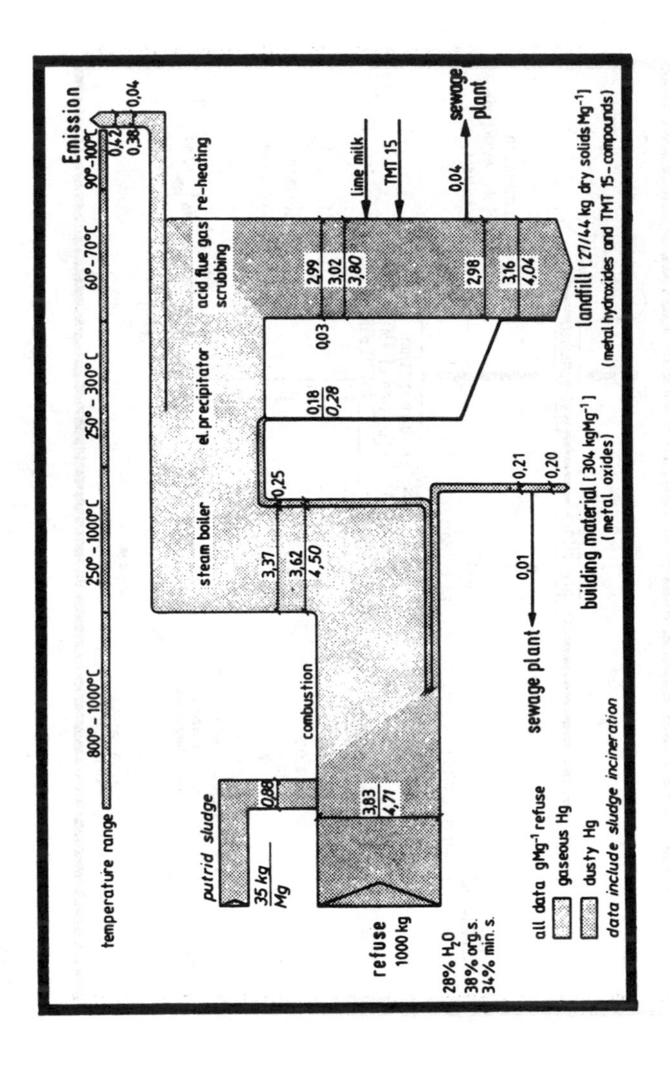
temperature range
800°-1000°C
250°-1000°C
250°-300°C
60°-70°C
90°-100°C
Emission
0,42
0,38
0,04
putrid sludge
35 kg/Mg
0,88
refuse
1000 kg
28% H2O
38% org.s.
34% min. s.
3,83
4,71
combustion
steam boiler
3,37
3,62
4,50
0,25
el. precipitator
0,18
0,28
0,03
acid flue gas scrubbing
re-heating
2,99
3,02
3,80
lime milk
TMT 15
0,04
sewage plant
2,98
3,16
4,04
landfill [27/44 kg dry solids Mg-1]
(metal hydroxides and TMT 15-compounds)
0,21
0,20
0,01
sewage plant
building material [304 kgMg-1]
(metal oxides)
all data gMg-1 refuse
gaseous Hg
dusty Hg
data include sludge incineration

Figure 2

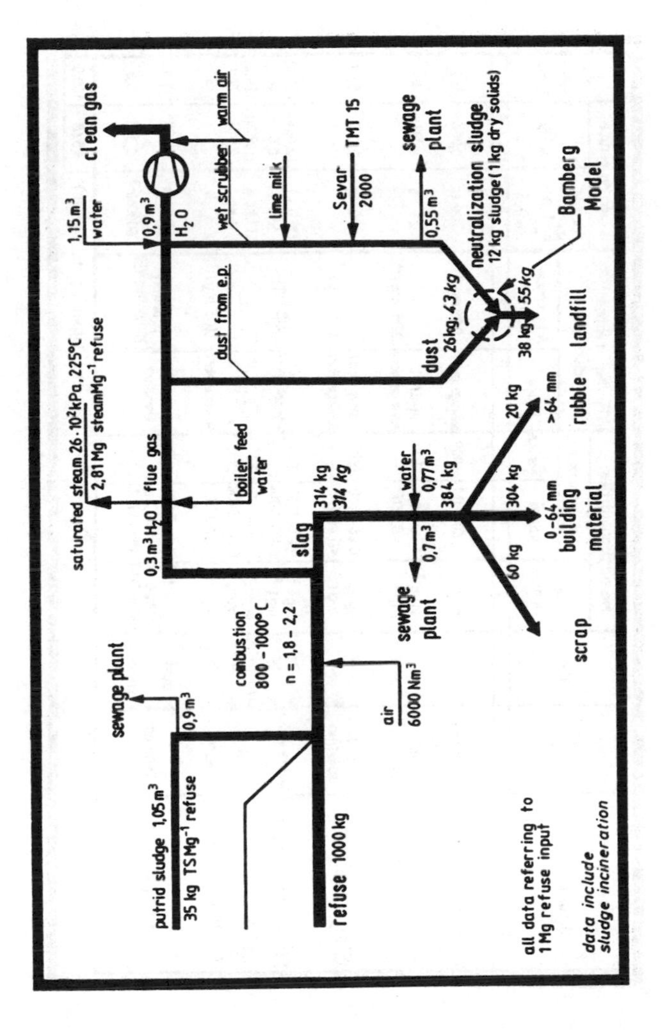

putrid sludge 1,05 m^3
35 kg TS Mg^{-1} refuse
sewage plant
0,9 m^3
refuse 1000 kg
air
6000 Nm^3
combustion
800 - 1000°C
n = 1,8 - 2,2
slag
0,3 m^3 H_2O
saturated steam 26 · 10^2 kPa, 225°C
2,81 Mg steam Mg^{-1} refuse
flue gas
boiler feed water
314 kg
314 kg
water
0,77 m^3
384 kg
0,7 m^3
sewage plant
60 kg
304 kg
20 kg
scrap
0 - 64 mm building material
> 64 mm rubble
dust from e.p.
dust
26 kg; 43 kg
38 kg
55 kg
landfill
H_2O
0,9 m^3
1,15 m^3 water
wet scrubber
lime milk
Sevar 2000
TMT 15
clean gas
warm air
0,55 m^3
sewage plant
neutralization sludge
12 kg sludge (1 kg dry solids)
Bamberg Model
all data referring to 1 Mg refuse input
data include sludge incineration

Figure 3

Table 1

Occurence	Transport medium	Other	Mercury quantities			Further treatment or discharge		per Mg refuse		% of Σ Hg
			Dimensions	Concentration						
				Range	Average	Place	Adm. concentration	Specific quantity of transport medium	g Hg	
1	2	3	4	5	6	7	8	9	10 = 6·9	11
Wet ash extractor	waste water	—	$mg\ l^{-1}$	0,007 - 0,1	0,009	sewage plant	0,05 $mg\ l^{-1}$	700 l	0,006	0,2
El. precipitator	dust	—	$mg\ kg^{-1}$	1,4 - 18,9	7,1	landfill	—	26 kg	0,184	4,7
	solution	—	*$mg\ l^{-1}$*	*<0,001*	—	*sewage plant*	*0,05 $mg\ l^{-1}$*			
Slag treatment	slag	—	$mg\ kg^{-1}$	0,46 - 1,80	0,70	constr. mat.	—	304 kg	0,213	5,7
	solution	—	*$mg\ l^{-1}$*	*<0,0005*	—	*sewage plant*	*0,05 $mg\ l^{-1}$*			
Flue gas scrubber after neutralization	waste water	w/o TMT15	$mg\ l^{-1}$	0,47 - 8,00	2,47	sewage plant	0,05 $mg\ l^{-1}$	(850 l)	(2,084)	(54,4)
		w TMT15	$mg\ l^{-1}$	0,003 - 0,094	0,048			550 l	0,026	0,7
	sludge	w/o TMT15	$mg\ kg^{-1}$	—	927	landfill	—	(1 kg)	(0,927)	(24,2)
		w TMT15	$mg\ kg^{-1}$	1790 - 3400	2985			1 kg	2,985	77,9
	solution	—	*$mg\ l^{-1}$*	*<0,005*	—	*sewage plant*	*0,05 $mg\ l^{-1}$*			
Stack	clean gas	dusty	$\mu g\ (Nm^3 f)^{-1}$ 11% O_2	1,4 - 10,8	6,2	emission	probably 200 $\mu g\ (Nm^3 f)^{-1}$	6000 Nm^3	0,037	1,0
		gaseous	$\mu g\ (Nm^3 f)^{-1}$	48,6 - 76,8	62,8			6000 Nm^3	0,377	9,8
							Σ Hg Output ($g\ Mg^{-1}$ refuse)		3,828	100,0

3.3 Hg in Slag and Filter Dust

Only very low concentrations of mercury are present in the slag, which is intended for use as a construction material as well as in the slag washing water (18). This results from the high temperatures, to which the incineration material is subjected. If, in exceptional cases, higher concentrations are bound in these materials, this is due to insufficient incineration, high temperature resistance, e.g. for HgS with corresponding vapor pressure, or to the fact that the incineration temperatures are too low. The percentage of mercury in the furnace- and filter-dust is also very low, because with the flue gas purification procedure the electrostatic precipitator (e.p.) temperature varies between 200 - 300° and the condensation temperature of Hg compounds is in this range. Already condensed Hg compounds are present as solid material in the form of microparticles, which can only be separated in the e.p. to a limited degree. The binding of the fine particles in the aqueous phase is assumed by the subsequent flue gas wet scrubber.

If the temperature for removal of the dust in the flue gas is lowered to approx. 150°C as is common in the semi-dry procedure, the Hg content in the occurring sorption material increases highly. However in addition to temperature reduction and the associated condensation of Hg and its compounds in these flue gas cleaning processes, the addition of lime, for example, with its large surface plays a significant role in binding the acidic pollutants to scavenge Hg. Assuming a dry flue gas quantity of 5000 - 6000 Nm^3 mg^{-1} garbage following the e.p. a quantity of approx. 0.6 - 0.5 mg Hg Nm^{-3} results without digested sludge incineration. This high mercury emission value is frequently still used today for generalized comparison of the purity of energy obtained from garbage and from other primary energy sources, although it applies only for plants without further flue gas purification. Opponents of garbage incineration occasionally use this erroneous information to create prejudice and fear on the part of ecologically minded citizens against garbage incineration plants.

With the flue gas purification using the dry and semi-dry procedures total gaseous mercury quantities of 0.04 - 0.57 mg Nm^{-3} are present in the purified gas (19, 20, 21). The salts from flue gas purification have considerably higher Hg concentrations compared to the wet process. The large range of variation for the specified emissions is certainly accountable to the lower condensation effect with these methods in comparison to wet washing as well as to the various concentrations in the initial material. Dust removal processes in the unpurified gas should be accomplished above the dew point, in order to prevent clogging, formation of lumps and incrustation in the dust remove equipment, whereby a differentiation must be made between the acid and water dew point.

3.4 Mercury in the Flue Gas Washings

In the following description wet washing is focused, because it is presently the method by which the most efficient elemination of Hg can be achieved (23, 24). Figure 3 and table 1 show the washing effect with the usual pH value of 0.5 - 1.0 in the flue gas washing water. This low pH value results from washing out the acidic pollutant in the flue gas and therefore provides ideal prerequisites for keeping heavy metal salt in solution.

3.4.1 Theories on Hg Binding via Flue Gas Washing

Among specialists various opinions still exist regarding the exact relationships for washing Hg out of the flue gas, whereby only a few aspects are quickly mentioned here:

a) for all theories agreement exists that the Hg condensation can be influenced decisively by the greatest possible temperature reduction,

b) the efficient removal of Hg by acidic flue gas washing is not questioned on the basis of the achieved and measured Hg concentrations,

c) a high degree of uncertainty exists whether Hg is present in the flue gas washing water in metallic form or in the form of dissolved Hg salts or oxides, whereby the possibility of the presence of metallic mercury in the flue gas is classified as slight by experts,

d) there is still no clear opinion whether the very stable and easily soluble Hg (II) chloro- complex $Hg_2[HgCl_4]$ is formed by a flue gas washing water temperature of approx. 70°C and by the presence of sulphuric acid with high surplus chloride together with Hg compounds and possibly even metallic Hg (1, 25).

3.4.2 Removal of the Mercury in the Flue Gas Washing Water

In addition to an average chloride content of 10 - 20 g l^{-1} the acidic waste water from flue gas washing (in Bamberg approx. 0.5 m^3 Mg^{-1} garbage) contains an average Hg-content of 3 - 4 mg l^{-1}, which can increase to over 15 mg l^{-1}. Highly deviating data on the previous concentrations is attributable less to the differences in the pollution contained in the garbage than to the dilution or concentrating effect resulting from addition of various processing water quantities to the flue gas washings. The flue gas washing water quantities can vary between 0.3 to over 1.0 m^3 Mg^{-1} garbage depending upon the operational procedure, the water and waste water requirements as well as the costs.

Within the scope of final neutralization, for example with milk of lime $Ca(OH)_2$, most of the problematic heavy metals are precipitated to more than 80 - 90 % to far below the permissible discharge concentrations in the form of hydroxides. For cadmium and nickel the achievable precipitation degrees of approx. 70 % are sufficient to satisfy the discharge conditions applicable today. Problems result in elimination of the Hg. With neutralization using milk of lime the reaction from $Hg_2(HgCl_4)$ to Hg (II) oxide is incomplete due to the high surplus chloride (26). Long-term tests in Bamberg on the achievable Hg precipitation results through neutralization alone showed precipitation degrees of 30 - 40 % (24). In this case the residual concentration of Hg was approx. 2.5 mg l^{-1} and higher and exceeds the permissible discharge value by several times. In contrast to sulphide precipitation or ion exchanger the economical, reliable and harmless precipitation of mercury with the sodium salt of trimercapto-s-triazine $(C_3N_3S_3)^{3-}$, TMT 15 (Degussa, Frankfurt, West Germany) or SEVAR 2000 as a combined Hg precipitation and flocculation agent is available today for improvement of the mercury binding. Capital and operating costs occur through addition of slight quantities of these precipitation agents (approx. 50 - 100 ml m^{-3} flue gas waste water) depending upon the quantity of mercury and other heavy metals to be precipitated amounting to approx. 0.25 - 0.50 DM Mg^{-1} garbage (24).

The degree of mercury precipitation can be increased to 99 % and more with this method. The discesive Hg (II)-TMT compound is very difficult to dissolve, resistant to acids and temperature and therefore has hardly measurable leaching characteristics. The precipitation degree for cadmium and nickel is increased highly as a side effect by TMT precipitation.

3.4.3 Hg Contaminated Oxide Hydrate and TMT-Sludges

The hydroxide and TMT sludge with its low salt content due to the washing effect (26) occurs in quantities of approx. 1 kg dm Mg^{-1}. Freshly precipitated the sludge contains approx. 5 - 6 % dm. A compact, thick suldge with approx. 20 % dm can be created through mechanical dehydration. The Hg content is concentrated to approx. 3 g kg^{-1} dm and higher equal to $\geq$ 0.3 %.

Since depositing of this compact, however still mushy neutralization sludge presents problems, a material was required for compacting this oxide hydrate sludge. A material found as a compactor and simultaneously a product to be disposed of was the el. precipitator dust, which first had to be prepared for depositing by mixing with water to form a dust-free substance.

A new, compactable and only slightly permeable mixed product, which was difficult to leach (27), resulted from forced mixing of the hydroxide sludge from the flue gas washings and the filter dust. This procedure is called the BAMBERG MODEL (28). The product created in this manner offers an optimum sealing layer for domestic garbage depots in thicknesses of 1 m and more.

Long-term leaching tests on the mixed product at the Bamberger incineration plant with pH-values between 2.5 and 7.0 have confirmed the previous experience that Hg can only be leached to an extremely low degree in an acidic environment. This applies even after decomposition of the alkaline buffer effect of the mixed product, which forms as a result of the alkaline e.p. dust and the neutral sludge. Hg in the seepage water from the residual materials from garbage incineration does not lead to a problem at domestic garbage or mono-landfill, because it is far below the permissible concentrations in the waste water.

3.5 Hg Emissions from the Smokestack

The pollutions remaining in the purified gas can be in the form of dust or gases. When analyzing the total Hg emissions particular attention must be paid to the gaseous constituents. Acceptable results for the gaseous Hg constituents can be achieved with the Impinger procedure (29, 30). Care must be taken in selecting the proper, representative sampling location, taking the samples over a number of hours with a sufficient quantity of gas and good filtration of fine dust. The finest particles can penetrate through the filter under unfavorable conditions and lead to erroneous results with excessively high gaseous pollutant constituents. The test results for several plants with unobjectionable filtration are listed below in relationship to the flue gas purification procedure.

Dry procedure with subsequent cloth filter (19)

Hg in form of dust	Gaseous Hg
4×10^{-5} mg Nm^{-3}	0.22 - 0.57 mg Nm^{-3}
2.3×10^{-4} mg NM^{-3}	0.12 - 0.26 mg Nm^{-3}
4.3×10^{-4} mg Nm^{-3}	0.15 - 0.21 mg Nm^{-3}

Semy-dry procedure with preceeding cyclone and subsequent el. precipitator (22)

$< 10^{-5}$ mg Nm^{-3}	0.005 - 0.20 mg Nm^{-3}

Wet method with preceeding el. precipitator (18, 31, 32, 33)

4×10^{-4} mg Nm^{-3})) 6×10^{-3} mg Nm^{-3})	0.05 - 0.08 mg Nm^{-3}

Depending upon the selected flue gas purification procedure the ratio of gaseous mercury to mercury in dust form in the purified gas is $1:10^{-1}$ to $1:10^{-4}$. The gas phase is preeminent to such a degree that only its percentage is of significance in the consideration of the emissions. Further reduction of the mercury in dust form with additional dust removal systems does not lead to any mentionable improvement of the emission value.

Good binding of the gaseous percentage of Hg in the wet procedure is accountable to the condensation connected with temperature reduction. The comparably favorable emission value for Hg using the semi-dry procedure may by explained by the cooling effect - and the connected condensation effect - resulting from the addition of milk of lime. The gaseous mercury can be bound by the lime due to the evaporation the milk of lime sprayed in and the simultaneous availability of a very large condensation and absorption surface. The deposition of gaseous mercury in the precoat layer of the cloth filter attributed to the dry procedure does not appear to be confirmed.

4. Endangerment of Employees at Garbage Incineration Power Plants from Mercury

As a matter of principle it can be stated that employees at garbage incineration plants are not endangered any more than the average population by contact with materials and gases containing mercury in the prescribed concentrations. In this regard Hg examinations were performed on employees during two subsequent years in the form of blood and urine tests, whereby the mercury values were found to be approx. 5 - 50 % above the upper standard limit (see chapter 2.3). The median for the entire group of garbage incineration employees of 20 % of the upper standard limit was slightly higher than the median value for the examined population group, which was not subjected to the possible Hg-exposure in a garbage incineration plant (34).

5. Hg Reduction in Garbage for Emission Reduction

The Hg content in the incoming garbage is and remains decisive for the level of Hg-concentration in the various residual materials and in the emission. In addition to the only slightly controllable Hg content in domestic garbage, which consists primarily of Hg in vegetables, particularly the industry and the use of industrial products is responsible for the major percentage of mercury pollution. Examples of areas of applications containing Hg are chemicals, insecticides, dental amalgam and disinfectants used in medicine. Fungicides in paints, catalysts, thermometers, barometers, electronic components, fluorescent tubes, and similar lead to considerable quantities of mercury waste (35). More than 50 % of the total mercury content in garbage is attributable to used batteries. The high degree of pollution is not attributable primarily to mercury oxide batteries but rather to alkaline manganese batteries, which are only separated today in the rarest cases (36). The most effective method to reduce the quantity of mercury in untreated garbage would be to focus on collecting used batteries separately.

Literaturverzeichnis / Literature

(1) Ullmanns, 1980, "Enzyklopädie der technischen Chemie", Band 19, 4. Auflage, S. 643-671
Verlag Chemie, Weinheim

(2) Tabellenbuch Chemie, 1966, S. 37
VEB Deutscher Verlag für Grundstoffindustrie, Leipzig

(3) Christen, 1969
"Grundlagen der allgemeinen und anorganischen Chemie"
2. Auflage, S. 500-503, Sauerländer AG, Aarau

(4) Hofmann-Jander, 1972
"Qualitative Analyse"
Göschen Band 2619, 4. Auflage
Walter de Gruyter

(5) ABC Chemie, 1965, S. 1157 ff
Verlag Harri Deutsch, Frankfurt

(6) Römpps "Chemie-Lexikon", 1975
7. Auflage
Franckh'sche Verlagshandlung, Stuttgart

(7) Welz, Melchen, 1984, "Picotrace Determination of Mercury using the Amalgation Technique"
Atomic Spectroscopy Vol. 5, No. 2, March-April p. 38

(8) DIN 38406

(9) Regelwerk der Abwassertechnischen Vereinigung (ATV), 1983
"Hinweise für das Einleiten von Abwasser in eine öffentliche Abwasseranlage", Arbeitsblatt A 115 vom Januar

(10) Abwasserabgabengesetz -AbwAG- vom 1. Januar 1978

(11) Klärschlammverordnung -AbfKlärV- vom 1. April 1983

(12) Deutsche Forschungsgesellschaft (DFG), Senatskommission
"Aufstellung von Grenzwerten im biologischen Material"
Zentralinstitut für Arbeitsmedizin, Hamburg

(13) Schaller, K., Triebig, G., Valentin, G., 1983
"Praktische Hinweise für die Durchführung arbeits-medizinisch-toxikologischer Untersuchungen",
Arbeitsmedizin Aktuell, Lieferung 13. November, S. 55-69
Gustav Fischer Verlag, Stuttgart

(14) Deutsche Forschungsgesellschaft, 1981
"Quecksilber - Toxikologisch-arbeitsmedizinische Begründungen von MAK-Werten"
Verlag Chemie, Weinheim, 8. Lieferung

(15) Technische Anleitung zur Reinhaltung der Luft -TA-Luft,1986
Bundesimmissionsschutzgesetz (BImSchG) vom 15.3.1974/4.3.1982

(16) "Schweizerische Richtlinien über die Luftreinhaltung beim Verbrennen von Siedlungsabfällen" 1982
Eidgenössisches Department des Inneren vom 18.2.1982

(17) Reimann, D., 1984: "Belastung von Kläranlagen durch MVAs und/oder thermische Schlammkonditionierung"
Kommunalwirtschaft 5, S. 155-161

(18) Reimann, D., 1984: "Quecksilber bei der Müllverbrennung"
Umweltmagazin 6, S. 48-54

(19) Knorr, W., 1985: "Ergebnisse von Emmissionsmessungen an den Müllheizkraftwerken Kempten und Würzburg", KABV Saar-brücken, 5. Abfallwirtschaftliches Fachkolloquium 25./26.4.1985, S. 1-3

(20) Mosch Dr., H., Pfeiffer, K.: "Trockene Rauchgasreinigung in Müllverbrennungsanlagen", VGB-Heft TB 204, S. 165-182

(21) Cleve, V., 1985: "AFA-Verfahren zur trockenen Reinigung von Rauchgasen aus Müllverbrennungsanlagen", Technische Mitteilungen HdT., Vulkanverlag "Müllverbrennungsanlagen" 5, S. 236-241

(22) Horch Dr., K., 1984: "Rauchgasreinigung für die MVA München Nord", VGB-Heft TB 204, S. 181-198

(23) Erbach, G., Schöner, P., Maurer, P., 1984: "Schwermetallgehalt im Rauchgas hinter der sauren Wäsche", VGB-Heft TB 204, S. 140-164

(24) Reimann, D., 1984: "Reinigung von Rauchgaswaschwässern im MHKW Bamberg mit Schwerpunkt auf der Quecksilberliminierung durch TMT 15-Zugabe", VGB Kraftwerkstechnik 3, S. 230-235

(25) Degussa AG, 1985: "Hg-Untersuchungen", Unveröffentlichter Bericht des Geschäftsbereiches IC-Anwendungstechnik A, 6450 Hanau

(26) Reimann, D., 1984: "Chlorverbindungen im Müll und in der Müllverbrennung - Einfluss des Kunststoffes PVC", Müll und Abfall 6, S. 169-176

(27) Fichtel Dr., Beck, W., Giglberger, J., 1983: "Auslaugverhalten von Rückständen von Abfallverbrennungsanlagen - Rückstandsdeponie Großmehring - ", Schriftenreihe des Bay. Landesamtes für Umweltschutz, Heft 55

(28) Reimann, D., 1985: "Gemeinsame umweltschonende Beseitigung von Filterstäuben und Neutralisationsschlämmen aus der Rauchgaswäsche - BAMBERGER MODELL-, VGB-Heft TB 204, KABV Saarbrücken s. Abfallwirtschaftliches Fachkolloquium 25./26.4.1985, Technische Mitteilungen H.d.T., Vulkanverlag Müllverbrennungsanlagen, Heft 5, S. 268-272

(29) VDI 2452, Blatt 1

(30) "Empfehlungen über die Emissionsmessungen von Luftfremdstoffen bei stationären Anlagen", 1983, Anhang 14 und Anhang 25, Bundesamt für Umweltschutz, Schweiz

(31) Knorn Dr., Ch., Fürmaier, B., 1984: "Ergebnisse von Emissionsmessungen an Abfallverbrennungsanlagen", Müllverbrennung und Rauchgasreinigung, Herausgeber Thomé-Kozmiensky, Technik, Wirtschaft, Umweltschutz, Bd. 7 Müll und Abfall, S. 29-36

(32) Vogg Prof., H., 1985: "Emissionen aus Müllverbrennungsanlagen", VDI-Seminar BW 433201

(33) Vicard, J.F., Knoche Dr., M., 1985: "Die nasse Rauchgasreinigung hinter Müllverbrennungsanlagen" Vortrag s. Abfallwirtschaftliches Fachkolloquium, Saarbrücken 25./26.4.1985, S. 1-10

(34) Reimann, D., Bloedner, Cl.-D., 1985: "Keine erhöhten Schwermetallgehalte (Pb, Hg,Cd) im Blut des Betriebspersonals eines Müllheizkraftwerkes", Müll und Abfall 3, S. 72-76

(35) Lorber Prof., K., 1983: "Die Zusammensetzung des Mülls und die durch Müllverbrennungsanlagen emittierten Schadstoffe", erschienen in Müllverbrennung und Rauchgasreinigung, Herausgeber Thomê-Kozmiensky, Technik, Wirtschaft, Umweltschutz Bd. 7, S. 559-594

(36) Genest, W., Reimann, D., 1985: "Abfallproblematik von Altbatterien", Müll und Abfall, Heft 7, S. 217-224

Control of Heavy Metal Discharge with Sodium Borohydride

By J.A. Ulman

MORTON THIOKOL INC, VENTRON PRODUCTS, 150 ANDOVER STREET, DANVERS, MASSACHUSETTS 01923, USA

Introduction

This presentation reviews the use of sodium borohydride ($NaBH_4$) for the removal and recovery of metal contaminants from industrial waste streams. Results of our R&D program to develop and implement effective processes for the removal of metals from metal finishing effluents will be described in detail. Particular attention will be focused on: (1) laboratory development of procedures for removing copper from printed wire board (PWB) industry waste solutions, and (2) scaleup and implementation of these procedures at industrial waste treatment facilities.

Sodium Borohydride Chemistry

Since its synthesis approximately forty years ago, sodium borohydride has been shown to be a powerful and versatile reducing agent in numerous industrial processes. Its effectiveness in reducing metal cations has led to its application in the following areas: (1) removal of metals from waste streams to meet strict discharge limits;[1] (2) autocatalytic (electroless) plating of metals such as nickel,[2] copper,[3] and gold[4] on a wide variety of substrates; (3) recovery of precious metals from waste streams;[5] and (4) preparation of catalytic metal species used in the hydrogenation of organic compounds and other chemical processes.[6]

The effectiveness of sodium borohydride is due, in part, to its low molecular weight (37.8) and the ability of each molecule to deliver eight electrons to reducible metal cations. The reductions can be described by the following equations:

$$NaBH_4 + 8OH^- \longrightarrow NaBO_2 + 6H_2O + 8e^-$$

$$4M^{2+} + 8e^- \longrightarrow 4M^\circ$$

$$4M^{2+} + NaBH_4 + 8OH^- \longrightarrow 4M^\circ + NaBO_2 + 6H_2O$$

The reduction mechanism involves intermediate formation of an unstable metal hydride or borohydride. Several research groups have reported the isolation of copper and nickel borohydrides by carrying out the reaction in the presence of stabilizing ligands, such as certain phosphines.[7] In a recently completed academic study funded by Ventron, copper hydride (CuH) was detected when copper(II) was reduced by sodium borohydride in aqueous solution under inert atmosphere conditions.[8] Under conditions normally found in waste treatment operations, these intermediates decompose rapidly to a stable reduced metal species (and/or function as reducing agents for other metal cations). The particular metal species that is produced is dependent upon the metal ion. $NaBH_4$ can reduce metal ions to a lower oxidation state [e.g., $Fe(3+) \longrightarrow Fe(2+)$ or $Cr(6+)$

——> $Cr(3+)$], to the pure metal [e.g., $Pb(2+)$ ——> $Pb(0)$], to a volatile hydride [e.g., $Ge(4+)$ ——> GeH_4], or to an insoluble boride [e.g., $Ni(2+)$ ——> Ni_2B].

Sodium borohydride can reduce a large number of metal ions. A comprehensive bibliography on the subject that we completed recently lists references to the reductions of 34 different metals, some from more than one oxidation state.[9] The data in Table 1, which lists the metals most commonly encountered in metal removal and recovery, illustrate the reducing capability of sodium borohydride. In practice, quantities greater than those theoretically required should be used, since sodium borohydride will react with other components in solution.

TABLE 1. THEORETICAL SODIUM BOROHYDRIDE USE LEVELS AND QUANTITIES OF METALS RECOVERED

METAL	OXIDATION STATE	SODIUM BOROHYDRIDE THEORETICAL USE LEVELS		METAL RECOVERY
		POWDER (G SBH/KG METAL)	VENMET (ML VENMET/KG METAL)	(KG METAL/KG SBH)
COPPER	Cu^{2+}	143	850	7
LEAD	Pb^{2+}	46	270	22
NICKEL	Ni^{2+}	167	1000	6
GOLD	Au^{3+}	72	430	14
SILVER	Ag^{+}	43	260	23
CADMIUM	Cd^{2+}	82	370	12
MERCURY	Hg^{2+}	48	280	21
PALLADIUM	Pd^{2+}	91	540	11
PLATINUM	Pt^{4+}	100	600	10
COBALT	Co^{2+}	167	1000	6
RHODIUM	Rh^{3+}	143	850	7
IRIDIUM	Ir^{4+}	100	600	10

TREATMENT LEVELS SHOWN ARE FOR 97% ACTIVE SODIUM BOROHYDRIDE POWDER AND VENMET™, A STABILIZED WATER SOLUTION OF 12% SODIUM BOROHYDRIDE AND 40% NAOH (BY WEIGHT).

<u>Waste Treatment in the Metal Finishing Industry</u>

In recent years in the USA, Environmental Protection Agency (EPA), state, and local regulations covering discharge of metals have become increasingly stringent. EPA-mandated limits for copper are 4.5 mg/L as a maximum for any one day and 2.7 mg/L maximum for a four day average. In many cases, state and local requirements are even more strict. Effective methods of treating these solutions have become an important concern of metal finishers. In our contacts with industry personnel, we have found the following to be some of their criteria for choosing a treatment method:

(1) <u>Meeting discharge requirements</u> - The method must be capable of consistently removing metal concentrations to below required levels. This is not a trivial problem, since there is a great deal of variation in waste solutions even within a single treatment facility. Rinse solutions, acid and alkaline strippers, and spent electroless plating baths contain different concentrations of metals, different complexing agents, and different additives and impurities from the finishing process. Furthermore, there are day-to-day variations in the makeup of waste streams.

(2) <u>Favorable economics</u> - The total cost of the method (chemicals, capital equipment, operating expenses, and sludge disposal) must be competitive.

(3) <u>Sludge volume and metals content</u> - The handling costs and liabilities associated with shipping and

disposal of hazardous sludges have made sludge volume reduction a goal of many waste treatment operations. Production of high total solids, high metals content sludges is also attractive, since it offers the possibility of recovering metal values.

Laboratory Program on Copper Removal from PWB Industry Waste Streams

During the last few years we have been conducting an intensive program to develop effective processes for the treatment of printed wire board (PWB) industry wastes by chemical reduction with sodium borohydride. This program could be justified on the following grounds:

(1) There presently existed several established industrial processes employing sodium borohydride for the removal and/or recovery of metals from effluent streams: mercury from chloroalkali plants, lead from tetraalkyllead production, and silver and cadmium from photographic film manufacture.(10)

(2) As noted above, sodium borohydride has the ability to reduce many metal ions, including the majority of those found in industrial waste streams.

I shall describe the progress we made during the course of our program, from the early evaluations through the optimization of treatment procedures for a wide variety of waste streams. Results of plant trials and chemical and engineering aspects of the installation and operation of our system at a typical waste treatment facility will be discussed in the following section.

In the early stages of our program, we evaluated the effectiveness of sodium borohydride for the reduction and removal of copper (2+) ions from an alkaline etchant solution (ammoniacal cupric chloride), and acid etchant solution (copper-ammonium persulfate), a spent electroless copper bath, and chelated rinse water.

The alkaline etchant solution was used to remove unwanted copper from sections of the board after the circuit pattern was printed. It contained a very high concentration (20,000 mg Cu^{2+}/L) of $Cu(NH_3)_4^{2+}$ ions. The solution pH was 8.5-9.0. Since sodium borohydride reduction of metal ions is most effective in the neutral to mildly alkalinerange, no pH adjustment was necessary. The solution was diluted to facilitate later liquid-solid separation. VenMet™ (a stable, aqueous alkaline solution containing 12% sodium borohydride) was added dropwise, causing rapid reduction of copper. The reaction appeared to go through two stages. During the first stage, continued addition of VenMet™ gave a brown, turbid mixture as copper precipitated from solution. After about five minutes the reduction rate increased markedly, and large black particles formed. When agitation was stopped, the particles settled rapidly and were separated from the solution by decantation. The solution contained less than 0.5 mg Cu^{2+}/L, well under the allowable limits for discharge.

The spent electroless copper bath contained $Cu(EDTA)^{2-}$, a more strongly complexed form of copper than was found in the alkaline etchant. Due to the lower copper concentration (1800 mg/L), the sample was not diluted before treatment. The initial pH was 11, so no adjustment was needed.[11] We found a diluted (5:1 with water) form of VenMet™ to be most convenient for treating this less concentrated solution. The reduction process and final levels were as described for the alkaline etchant. However, solids settling was not as complete, so a final filtration was necessary.

The acid etchant solution had about the same copper concentration (1300 mg/L) as the spent electroless copper bath, but differed from it and the alkaline etchant in that strong oxidizers were present and the pH was very low. The oxidizers would consume large quantities of sodium borohydride, making the process uneconomical. The procedure that we developed uses the less expensive reducing agent sodium bisulfite to scavenge the oxidizers, followed by sodium borohydride (after caustic addition to raise the pH to between 6 and 8) to reduce copper ions. The reduction process and final copper levels were as described for the alkaline etchant. Settling of reduced particles was rapid, but filtration was necessary to remove small amounts of suspended fines.

Analysis of the results of these tests showed sodium borohydride reduction to be a promising technology for treating PWB industry wastes. Compared with conventional technologies, some advantages that we identified were:

(1) $NaBH_4$ reduced copper from complexed solutions to levels below those required for discharge. These low levels cannot be achieved by lime or caustic addition alone.

(2) Final sludges were compact. Volume reductions of tenfold or greater compared with the commonly used ferrous sulfate-hydroxide method for treating complexed wastes were obtained.

(3) Copper content of the sludges was high: typically greater than 80% (dry weight basis) in these tests. This could lead treatment facilities to consider recycling of the sludges for their copper values as an economic alternative to disposal. This option would be attractive also because it would remove or minimize their liability for hazardous waste.

(4) In these tests, a sharp drop in oxidation-reduction potential (ORP) occurred when reduction took place. Upon completion of the reduction, the ORP value was in the -800 to -1000 mV range. Thus, reaction control by ORP appeared feasible (see Figure 1).

Although the results obtained in this and other early studies were cause for optimism, several areas where improvements were needed were identified:

(1) When kept in contact with the solution, the reduced precipitate slowly reoxidized, causing metal concentrations to increase to unacceptably high levels.

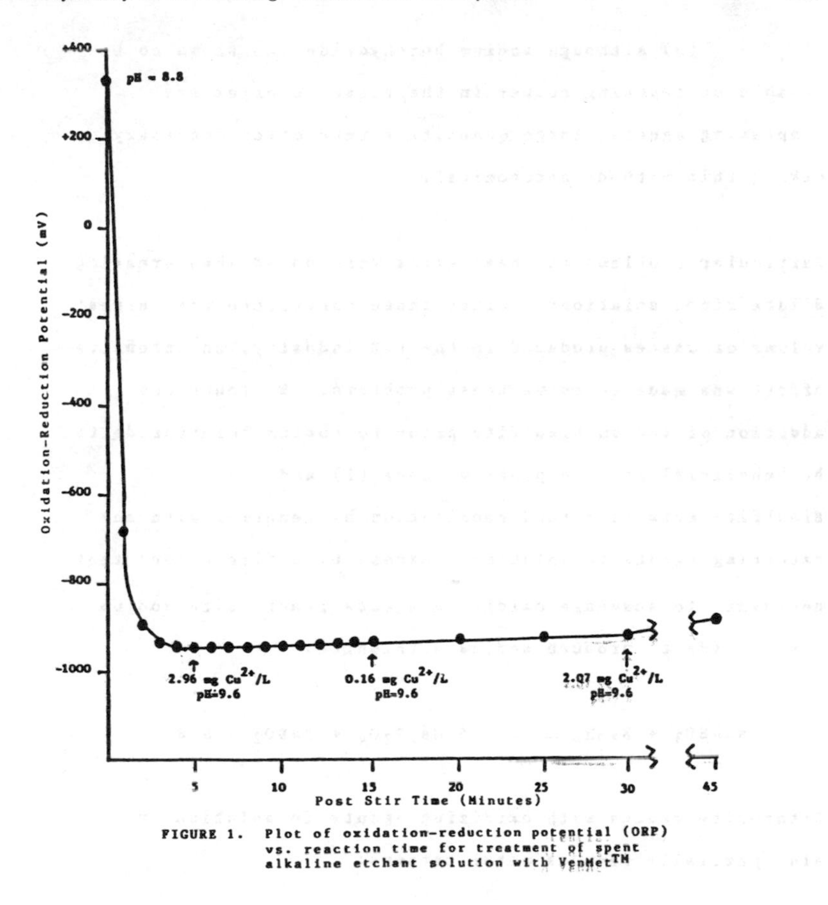

FIGURE 1. Plot of oxidation-reduction potential (ORP) vs. reaction time for treatment of spent alkaline etchant solution with VenMet™

The rate of reoxidation was much higher in the presence of complexing agents.

(2) In some instances, reduced particles were very fine, making solid/liquid separation difficult. Fine particles also reoxidized (redissolved) rapidly.

(3) Although sodium borohydride was shown to be capable of reducing copper in the presence of excess complexing agents, large quantities were often necessary, making this method uneconomical.

Particular problems in these areas were noted when treating dilute rinse solutions. Since those constitute the largest volume of wastes produced in the PWB industry, an intensive effort was made to solve these problems. We found the addition of sodium bisulfite prior to sodium borohydride to be beneficial in both problem areas (1) and (3).
Bisulfite acts to retard reoxidation by reacting with any oxidizing agents in solution. Excess bisulfite beyond that necessary to scavenge oxidizing agents reacts with sodium borohydride to produce sodium dithionite:

$$8\ NaHSO_3 + NaBH_4 \longrightarrow 4\ Na_2S_2O_4 + NaBO_2 + 6\ H_2O$$

Dithionite reacts with oxidizing agents in solution and also partially reduces metal cations:

$$Na_2S_2O_4 + 2\ Cu^{2+} + 2\ H_2O \longrightarrow 2\ Cu^{+} + 2\ NaHSO_3 + 2\ H^{+}$$

This reaction regenerates bisulfite, which maintains a reducing environment.

While the mechanism of sodium borohydride reduction in these systems has not been fully elucidated, we believe that Cu(2+) is reduced through a Cu(1+) species to give Cu(0). As the Cu(2+) is reduced to Cu(+) by sodium borohydride and sodium dithionite, the bond between the copper and complexing ligand is weakened considerably. This allows rapid reduction by borohydride to metallic copper.

Thus, by scavenging oxidizing agents, maintaining a reducing environment, and producing sodium dithionite [which reduces strongly complexed $Cu(2+)$ to weakly complexed $Cu(1+)$], sodium bisulfite helps to alleviate the reoxidation problem and brings borohydride usage down to a more economical level.

In problem area (2), we found that in many cases a binary flocculation system promoted particle build-up. This system consisted of a cationic polymer, generally a poly(quaternary ammonium) compound, followed by an anionic polymer, usually a hydrolyzed polyacrylamide. With proper polymer dosage and mixing, we obtained good flocs, in most cases, that could be separated by either decantation or coarse filtration. In addition to improving on liquid/solid separation, increasing the particle size reduced the rate of reoxidation. At this time, the particular polymers that function best are being determined on a case-by-case basis. We will soon begin more in-depth studies to attempt to put polymer selection on a more scientific footing.

Another way to slow the reoxidation rate is to passivate the surface of the reduced metal particles. We have identified a series of passivating agents that have been very effective for this purpose. The most effective ones, along with their chemical structures, are shown below:

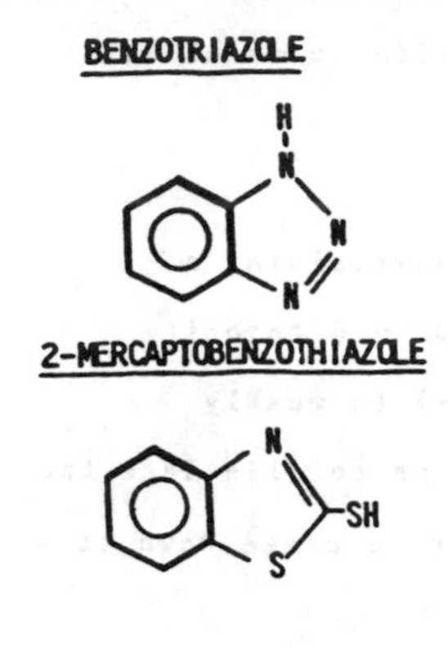

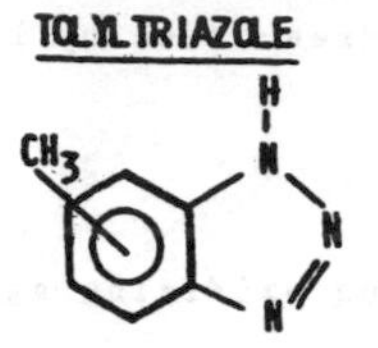

2-MERCAPTOBENZIMIDAZOLE

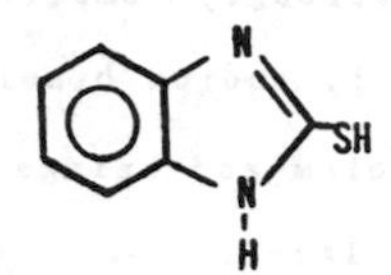

We believe these passivating agents function by forming a thin film over the reduced metal particles. This film, which probably consists of complex ions of metal and passivating agent, protects the metal from components in solution (e.g., chelating agent, oxygen) that would tend to redissolve it. The passivating agents are best added after the metals are removed from solution by reduction with sodium borohydride, but in certain cases can be added before reduction. Laboratory tests showed they could protect reduced copper, in the presence of ammonia and EDTA, from redissolving for periods of 24 hours or longer. A patent has been filed on the use of these passivating agents. We are now preparing to test them on a full plant scale.

In summary, as a result of extensive laboratory testing of PWB effluents, we developed the following general experimental procedure for treating rinses and more concentrated solutions.

(1) Add sodium bisulfite until the ORP stops decreasing appreciably. The pH should be maintained at 6-7 with caustic to avoid evolution of SO_2 gas.

(2) At pH 6-7, add VenMet™ until a particular ORP value is reached. This ORP value varies from one solution to the next, but is usually in the -600 to -800 mV range. The VenMet™ should be diluted (1:5, 1:10, or as convenient) with water before use.

(3) Continue stirring for approximately 15 minutes to allow reduction to go to completion.

(4) If necessary to improve solid/liquid separation, add cationic and anionic polymers in quantities determined by experiment. Quantities can range from 1 to 50 ppm.

(5) Allow sludge to settle, then decant (or filter) the supernatant.

At this point, we began field testing of this VenMet™ system at selected waste treatment facilities. The evaluation and optimization of the system at one typical facility will now be described in detail.

Case Study

This facility, at a major printed wire board manufacturer, already had a segregated treatment system in place for chelated and non-chelated wastes. Using a conventional ferrous sulfate ($FeSO_4$) - hydroxide system, they were able to meet discharge limits for copper and other regulated metals. However, sludge overloading was becoming a problem. The clarifiers and filter press were operating near capacity. The company wanted to increase production, but could not do so until this problem was solved. Since laboratory tests had demonstrated the ability of VenMet™ to reduce the sludge volume considerably compared with $FeSO_4$ hydroxide treatment, this promised to be an ideal test case for the VenMet™ system. This was also a test for the performance of VenMet™ in a continuous treatment system.At the start, the following requirements were established for evaluating the trial.

1. Sludge Volume Reduction - This was the primary goal. VenMet™ must be able to reduce complexed copper in all waste streams to a compact sludge that would allow for the handling of larger quantities of waste within the existing system capacity.

2. Effluent Compliance - Existing limits for copper (4.5 ppm daily maximum; 2.7 ppm maximum for four day average) must be met. The trial should also demonstrate the capability of VenMet™ to meet any future, more stringent EPA requirements.

3. Low Capital Costs - Modifications to existing equipment must be minimal.

4. <u>Economics</u> - Operating costs should be comparable to $FeSO_4$ system costs. Marginally increased operating costs were acceptable provided criteria (1)-(3) were met.

Figure 2

FERROUS SULFATE SYSTEM

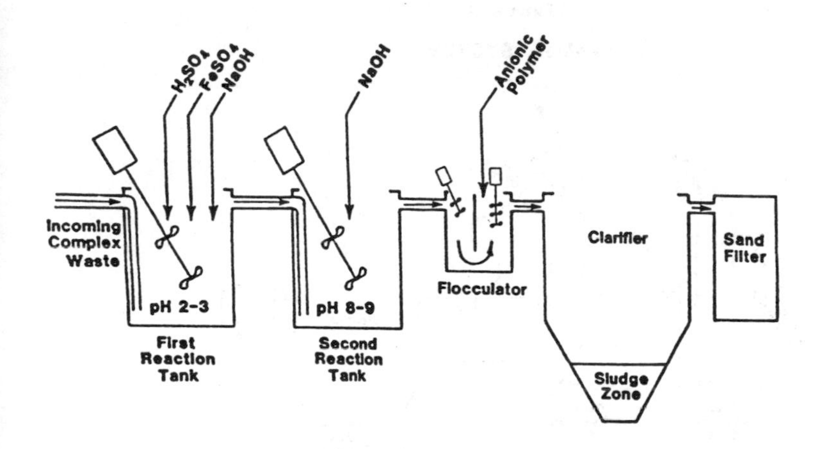

The company's existing waste treatment system, typical of many used in the PWB industry for chelated copper solutions, is shown in Figure 2. The process consisted of two reaction stages, a flocculation step, clarification, and final filtration. In the first stage reaction tankthe pH was controlled at 2-3 and 0.7% $FeSO_4$ solution was added, generally with a 14 to 1 molar ratio of iron to copper. The waste solutions then overflowed by gravity into the second tank. Here the pH was raised to 8.5-9.0 with 25% caustic solution, resulting in a ferric hydroxide-cupric hydroxide precipitate. Anionic polymer was added in a flash-mix stage immediately before the clarifier to improve flocculation.

Residence times for the different reaction steps were 20-40 minutes in the first and second tanks and 5-10 minutes in the flash-mix stage. The clarifier was sized for 1-1.5 hours of retention and dual, parallel, cross-linked sand filters removed any floc carryover from both the chelated and second non-chelated system. Sludge from the clarifiers was sent to a holding tank for further dewatering.

Figure 3

VenMet™ SYSTEM

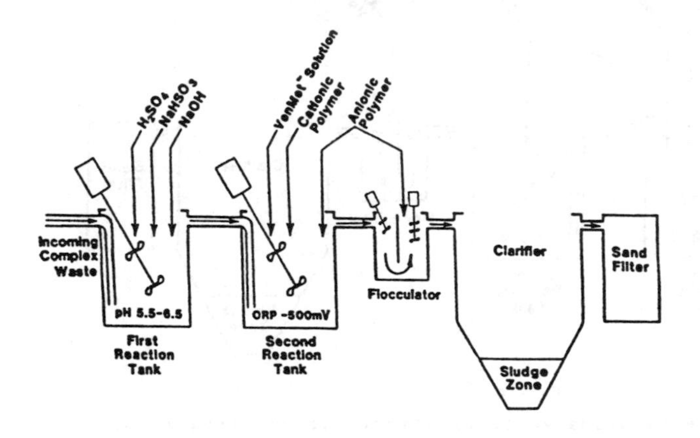

The conversion of the $FeSO_4$ system to the VenMet™ system was simple and straightforward (see Figure 3). The key engineering and process considerations were:

1. Adequate residence time for chemical reactions and flocculation. The 20-40 minute residence times in the first and second tanks were ideal to ensure complete copper reduction. Shorter times are usually adequate at lower pH's, while longer reaction times are needed at higher pH's.

2. Thorough mixing.

3. Accurate and stable process control (both pH and ORP). Control of the reaction pH is critical for effective and consistent reduction, while ORP control is necessary for efficient VenMet™ and sodium bisulfite usage.

4. Good flocculation and settling. In general, a binary (cationic and anionic polymers) flocculation system is effective for good settling. In some cases a final filter, such as a sand filter, is necessary to ensure acceptable effluent quality.

Necessary modifications to the existing system were minimal. In the first reaction tank, the pH control point was changed from 2-3 to 6, and a 10% solution of sodium bisulfite was substituted for the $FeSO_4$ solution. The following changes were made on the second reaction tank. The pH controller was converted to an ORP controller for metering in dilute VenMet™ solution, the 25% caustic solution was replaced with 10% VenMet™, and a cationic polymer addition step was added.

During start-up and subsequent operation the performance of the VenMet™ system was further optimized:

1. The pH set point in the first reaction tank was lowered from 6.0 to 5.5. This increased the reaction rate and increased particle size. With good mixing, hydrolysis was not a problem.

2. The ORP set point in the second reaction tank was optimized at -500 mV. This allowed reduction to the required copper levels with minimum VenMet™ usage.

3. $NaHSO_3$ usage was optimized to maximize copper removal and minimize flocculation and settling problems.

4. Incoming copper levels were unusually low, typically below 20 ppm. Such dilute solutions decrease reduction efficiency and present settling problems. Performance was improved by bleeding concentrated wastes into the continuous systems at a constant, uniform rate to maintain minimum copper levels above 20 ppm.

5. The anionic polymer addition point was moved from the flash-mix stage to the second reaction tank to provide adequate mixing time.

After the VenMet™ system had operated for sixty days, the total costs in chemicals and sludge handling were calculated and compared with the ferrous sulfate system. The cost of treating the non-chelated waste, which is assumed to be constant, is included in both cases.

Ferrous Sulfate System	per 60 days	approximate cost
ferrous sulfate	20,000 lb	$2,300.00
sodium hydroxide	12,000 gal	4,000.00
sulfuric acid	400 gal	530.00
flocculants	15 lb	21.00
	Chemical cost per 60 days	$6,851.00
	Chemical cost per day	$ 114.00
	Sludge handling cost per day	$ 215.00
	Total Cost Per Day	$ 329.00

VenMet™ System	per 60 days	approximate cost
VenMet™ solution	186 gal	$6,000.00
sodium hydroxide	7,500 gal	2,400.00
sulfuric acid	330 gal	390.00
sodium bisulfite	13,000 lb	3,700.00
flocculants	27.7 gal	444.00
	Chemical cost per 60 days	$12,934.00
	Chemical cost per day	$215.00
	Sludge handling cost per day	$74.00
	Total Cost Per Day	$289.00

The VenMet™ system thus showed an overall cost decrease of 12% per day, or $10,000 per year. Sludge generation from the chelated waste stream was reduced by 82%, resulting in a combined (chelated plus non-chelated) sludge generation decrease of 66%. After dewatering, the filter cake contained 55% total solids, compared to 35% total solids average with ferrous sulfate treatment. The operation of the filter press was reduced from continuous operation to approximately three times per week. Cycle times were cut in half.

In addition to the above benefits, the copper content of the sludge from the chelated wastes increase from approximately 5% to 80% on a dry weight basis. The sludge is thus suitable for refining to recover copper values. One of the major refiners of copper ores in the United States is presently accepting sludges from this and other VenMet™ users. The refiner has determined that these sludges fit directly into their smelting operation. At the present time, the depressed value of copper metal in the United States has not made it economically worthwhile for the reclamation of copper values. However, the sludge is being utilized by the refiner in a different application. The great advantage to the waste treatment facility, in either case, is the elimination of their liability for disposal of the hazardous waste.

Summary

The development of the VenMet™ system for the treatment of printed wire board industry waste streams has been described in detail. The benefits foreseen for this system based on laboratory beaker studies have been realized in both continuous treatment system at this and other facilities, and in batch treatment systems. Copper (and other metals) concentrations have been reduced, in most chelated solutions, to low levels, allowing compliance with environmental agency regulations. The all-liquid VenMet™ system can be operated automatically with ORP control. Sludge volume reductions vs. conventional treatment methods have been significant. This allows for increased production rates and results in considerable savings in sludge disposal costs. And, because of their favorable characteristics, these sludges are being accepted by a refiner, thereby minimizing the waste treater's liability for hazardous waste disposal. The VenMet™ system is gaining widespread acceptance in the industry, and should continue to do so as we optimize all parameters of the system.

References

1. (a) M. J. Lindsay and M. E. Hackman, "Sodium Borohydride Reduces Hazardous Waste", presented at the Purdue Industrial Wastewater Conference (May 1985).

(b) M. E. Fleming and J. A. Ulman, "Sodium Borohydride Environmental Control Application: Reduction of Nickel(II) Complexes in Spent Electroless Plating Bath," presented at Electroless Nickel Conference IV, Chicago, April 1985.

(c) J. A. Ulman, "Control of Heavy Metal Discharge in the Printed Circuit Industry with Sodium Borohydride," presented at the 1984 AES SUR/FIN Annual Technical Conference and Exhibit, New York City (July 1984).

(d) S. F. Heleba, "EPA Effluent Compliance and Hazardous Sludge Control with Sodium Borohydride," presented at the Massachusetts Hazardous Waste Source Reduction Conference and Exhibition, Boxboro, MA (October 1983).

(e) R. N. Duncan and J. R. Zickgraf, Products Finishing, January 1982, 55.

(f) K. Parker, "The Waste Treatment of Spent Electroless Nickel Baths," presented at the First AES Electroless Plating Symposium, St. Louis (March 1982).

(g) F. S. Tuznik and A. A. Lis, Proc. World Congr. Met. Finish, 11th, 1984, 297.

2. (a) K. Lang, Electroplating Metal Finishing, 1966, 19, 86.

(b) H. Neiderpruem and H. G. Klein, Metal Finishing J., 1971, 17, 18.

(c) R. N. Duncan and T. L. Arney, Plating and Surface Finishing, 1984, 71, 49.

3. (a) R. J. Zeblisky, F. W. Schneble, and J. F. McCormack, U.S. Pat. 3,485,643 (1969).

(b) Y. Arisato and H. Koriyama, U.S. Pat. 4,138,267, (1979).

4. (a) P. Prost-Tournier and C. Allemmoz, U.S. Pat. 4,307,136 (1981).

(b) M. F. El-Shazly and K. D. Baker, U.S. Pat. 4,337,091 (1982) and references therein.

5. G. L. Medding and J. A. Lander, Precious Met. Proc. Abstr., 97,220285q(1982).

6. (a) R. C. Wade, D. G. Holah, A. N. Hughes, and B. C. Hui, Catal. Rev. -- Sci. Eng., 1976, 14(2), 211.

(b) R. C. Wade, Chem. Ind.(Dekker), 5 (Catal. Org. React.) (1981).

7. See, for example, D. G. Holah, A. N. Hughes, B. C. Hui, and K. Wright, Can. J. Chem., 1974, 52(17), 2990.

8. R. D. Archer, personal communication.

9. Morton Thiokol, Inc./Ventron Products, "Borohydride Reduction of Metal Cations. Abstracts of Patents and Literature."

10. M. M. Cook, J. A. Lander, and D. S. Littlehale, "Case Histories: Reviewing the Use of Sodium Borohydride for Control of Heavy Metal Discharge in Industrial Waste Waters", Proceedings of the Industrial Waste Conference, Purdue University, May 8-10, 1979; Arbor Science Publishers, Inc.; Michigan (1980).

11. Although sodium borohydride will reduce Cu^{2+} at pH 11, we later found reduction at lower pH to be preferred.

Determination of Trace Metals in Solution by Ion Chromatography

By R.A. Cochrane

DIONEX (UK) LTD, EELMOOR ROAD, FARNBOROUGH, HAMPSHIRE GU14 7QN, UK

The first published paper on Ion Chromatography appeared in September 1975. That paper (1), 'Novel Ion Exchange Chromatographic Method Using Conductive Detection' by H. Small, T.S. Stevens and W.C. Bauman of the Dow Chemical Company demonstrated for the first time the use of two Ion Exchange resins in series or dual column Ion Chromatography. The first column was used as the separator and the second column to suppress the eluent conductivity and thus demonstrated the use of a conductivity cell as a universal and very sensitive detector for ions.

Eleven years after its commercial introduction, Ion Chromatography has become one of the fastest growing areas in analytical chemistry. In that time the range and diversity of ions amenable to Ion Chromatographic analysis has increased rapidly from simple inorganic anions and cations such as fluoride, chloride, nitrite, phosphate, bromide, nitrate, sulphate, lithium, sodium, ammonium, potassium, magnesium and calcium to heavy metals, organics and amino-acids. This has all been made possible by the development of a variety of columns, new suppression systems and post column devices in combination with other detectors besides conductivity, such as electrochemical and optical detectors.

The analysis of metals by Ion Chromatography is a good example of how the diversity of the technique can be exploited to analyse the many forms that metals can take, for example, anionic, cationic or different charged states (speciation) and these will be used to describe Ion Chromatography more thoroughly.

SEPARATION TECHNOLOGY

The basic components of an Ion Chromatograph are shown in Figure 1. Separation is achieved by the relative affinities of ionic species for the functional groups of an ion exchange resin. These resins are based on an inert core of polystyrene-divinylbenzene which for anion exchangers are first surface sulphonated and then latexed via ionic bonding with a monolayer of anion-exchange particles, Figure 2. Cation resins are either simple surface sulphonated resins or as described previously a binary pellicular resin with cation exchangers attached to the surface of the resin spheres. The advantages of this type of resin are no pH limitations due to the inert core and the rigidity of the outer layer, which does not change size appreciably when converted from one ionic form to another. Mass transport rates on and off the functional sites are also much faster giving more efficiently separated components.

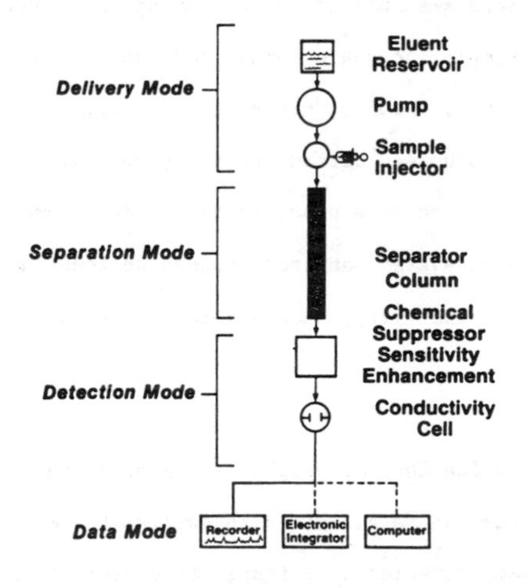

Figure 1

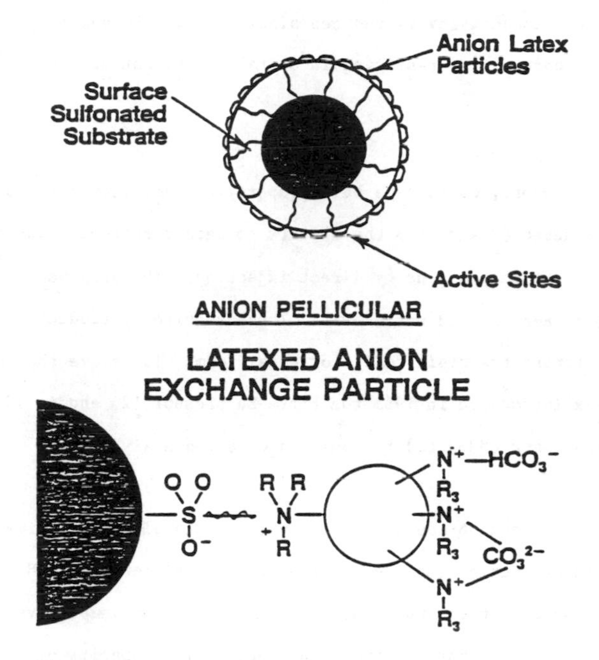

Figure 2

SUPPRESSION TECHNOLOGY

Detection by conductivity is enhanced by the use of a suppression system. In its original form this was a high capacity ion exchange resin which was able to convert a high conductivity salt to a low conducting acid or base. In the case of anion analysis the suppressor was a high capacity cation resin in the H+ form which would convert a sodium hydroxide eluent to water by the following mechanism:

$$\text{Resin-}\ H^{+} + Na^{+}OH^{-} \rightarrow \text{Resin-}\ Na^{+} + H^{+}O$$

At the same time sample anions are converted to their acid form.

$$\text{Resin-}\ H^{+} + Na^{+}\ A^{-} \rightarrow \text{Resin-}\ Na^{+} + H^{+}A^{-}$$

Thus the background conductivity is reduced significantly and the sample anions exhibit increased conductivity resulting in overall excellent sensitivity and high signal-to-noise.

The suppressor, therefore, converts a bulk property conductivity detector into a solute-specific detector and thus the ability to detect extremely low (parts per billion) levels of sample ions by direct injection. The original suppressors were columns packed with resin. These required periodic regeneration to restore the resin to its original form. To remove the need to regenerate, Dionex introduced in 1983 the Fibre Suppressor (2) and in 1985 the micromembrane suppressor (MMS) (3) for anion and cation analysis.

These were both based on membrane technology which would allow selective removal of the eluent counter ion and replacement with either H^{+} or OH^{-} regenerant for anion or cation analysis. The MMS is an improvement over the fibre suppressor because it has 100 times the suppression capacity of the fibre suppressor and a quarter the void volume (less than 50µl). The dynamic range of the MMS is 100 times better than non-chemically suppressed Ion Chromatography and 10 times better than the fibre suppressor.

Figure 3 shows an exploded view of the MMS and the alternating layers of high capacity ion exchange screens and ultrathin ion exchange membranes. Regenerant and eluent flow in opposite directions through the open sections of their screens. The screens provide a site-to-site pathway for eluent ions to the membrane and this is more efficient if the eluent stream has a membrane on either side. In an anion separating system with sodium hydroxide as a buffer the exchange reaction at the membranes is simply to replace the sodium with a hydrogen from the sulphuric acid regenerant.

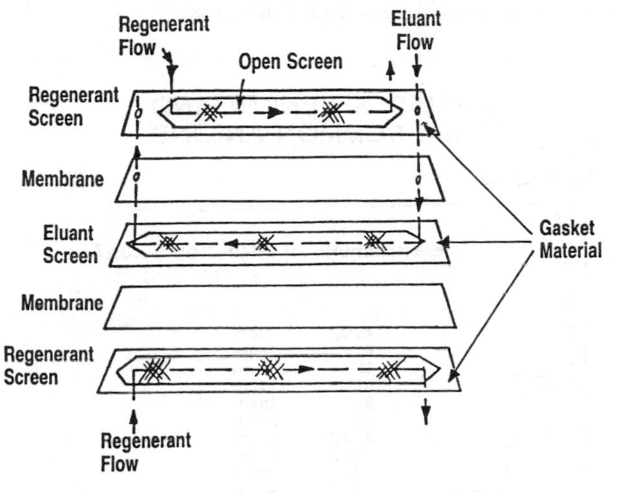

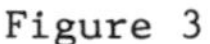

Figure 3

The high capacity of the MMS allows a greater choice of eluents and by permitting the use of much higher eluent concentrations gradient elution has been achieved. This significantly enhances the techniques utility as a methods development tool.

A comparison of all the suppression systems is shown in Figure 4.

Sensitivity Enhancement via Chemical Suppression

FEATURE	1975 PACKED BED	1983 FIBER	1985 MICROMEMBRANE
Continuous Regeneration	No	Yes	Yes
Sensitivity	Excellent	Excellent	Excellent
Capacity	High	Low	High (10-15 x Fiber)
Void Volume	2,000μL	200μL	45μL
Gradient Capability	No	No	Yes

Figure 4

Figures 5-13 show the range of oxy metal anions, metal complexes and common cations amenable to suppressed conductivity detection.

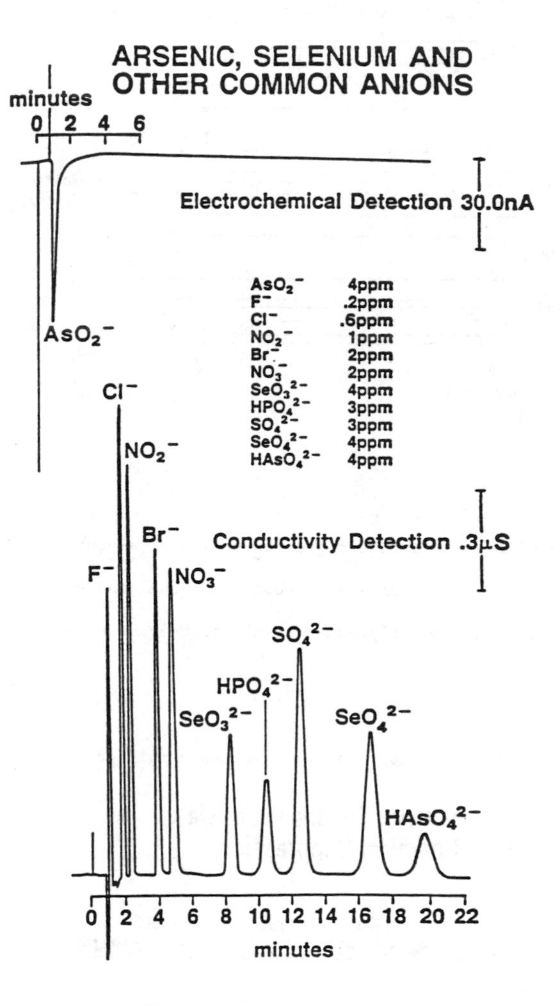

Figure 5. Arsenic and Selenium are important environmental pollutants, the different oxidation states being well separated from each other and the common anions. Arsenite is detected electrochemically.

Figure 6 shows the fast and sensitive detection of some of the important oxymetal ions, tungstate, molybdate and chromate.

Figure 7. By adding EDTA to an anion buffer it is possible to separate and detect the metal-EDTA anion complexes. This provides an alternative to the post column reaction technique to be described later.

Figure 8. Gold, Iron and Cobalt cyanide complexes can be separated and speciated on a polymeric reverse phase column with an ion pairing eluent that can be suppressed for conductivity detection. This is a powerful technique used to determine hydrophilic ionic compounds such as complexes and surfactants. This particular application is important for the plating industry because it is the only way of measuring gold (III) impurity in a gold plating bath.

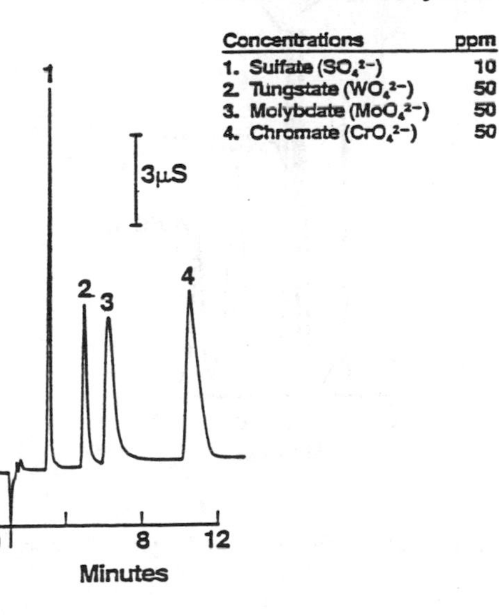

Figure 6

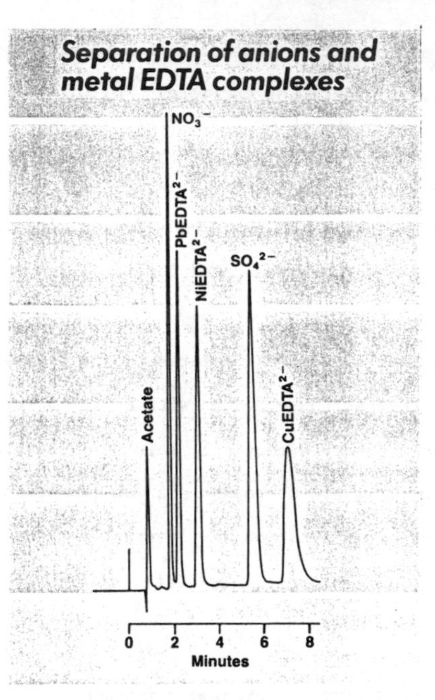

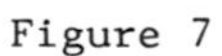

Figure 7

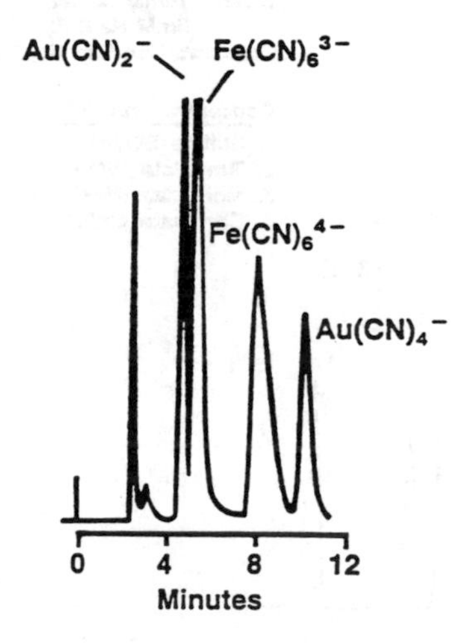

Figure 8 Separation of iron and gold cyanide complexes by MPIC

Figures 9 and 10 are examples of alkali and alkali earth metal separations on a pure cation resin. The same column can be used for either series, but it was not possible to separate them all in one run until a latexed cation resin was developed, Figures 11, 12 and 13.

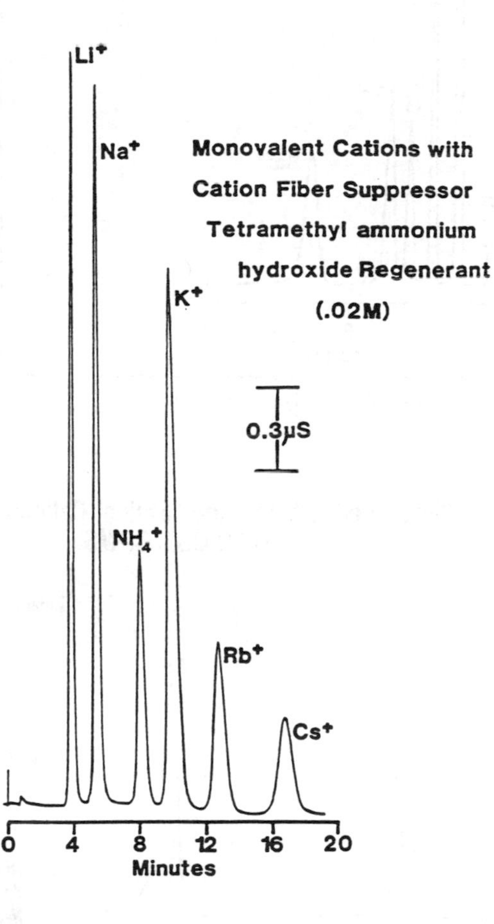

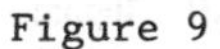

Figure 9

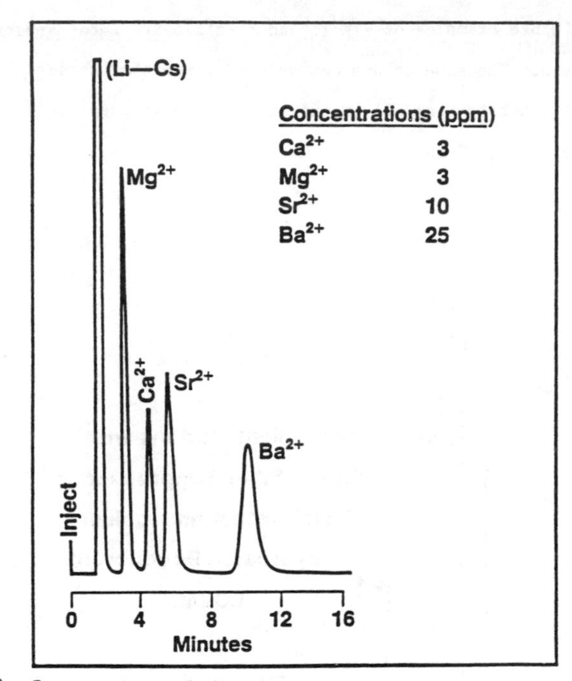

Figure 10 Separation of divalent cations

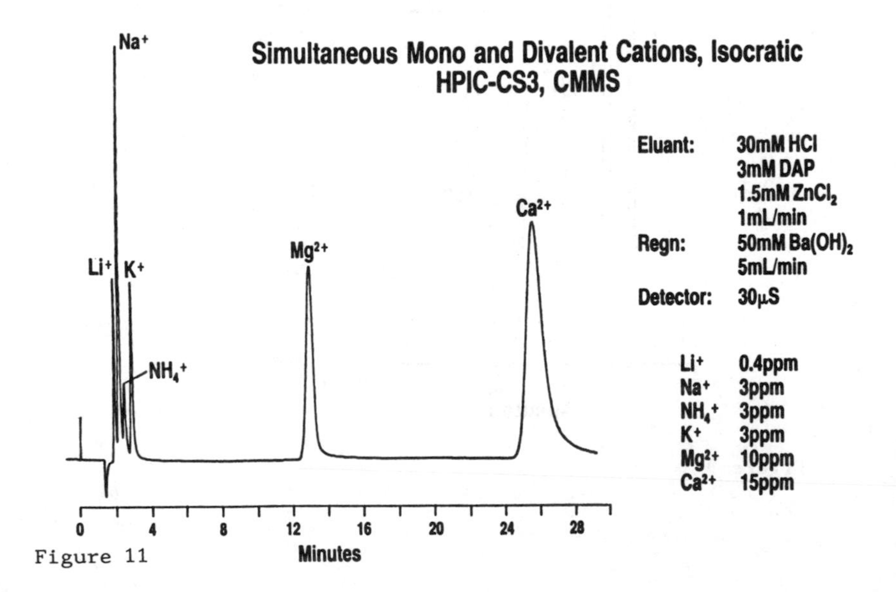

Figure 11

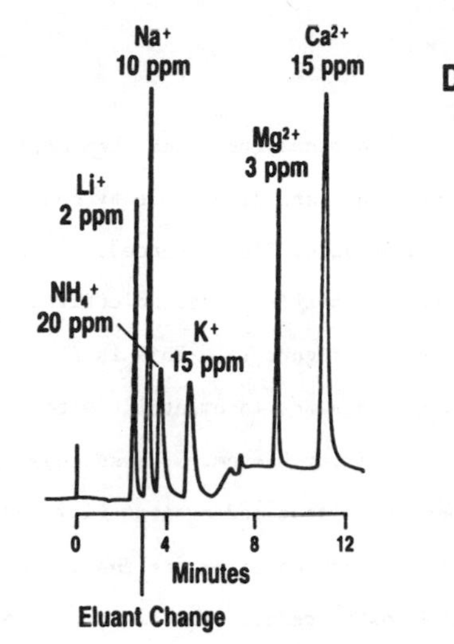

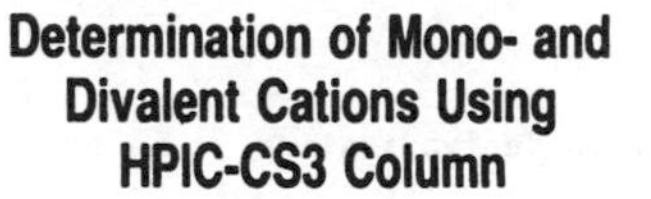

Determination of Mono- and
Divalent Cations Using
HPIC-CS3 Column

Figure 12

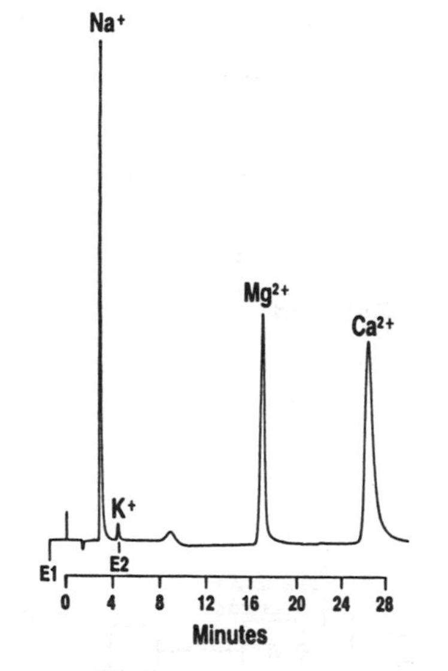

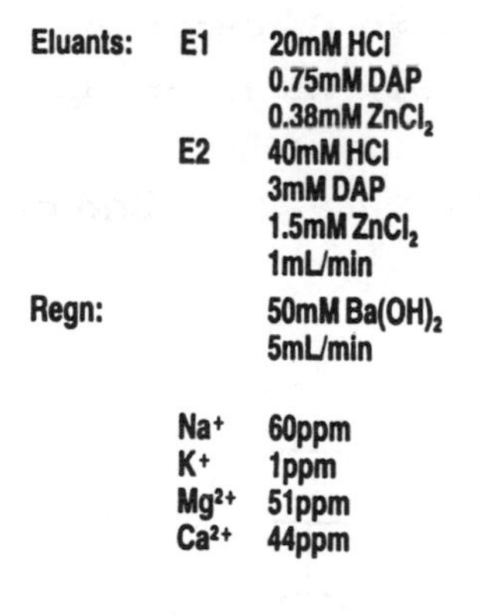

Figure 13

METALS BY POST COLUMN REACTION

Post column derivatisation is an established technique which chemically modifies the separated components so that they are detectable by one of the common chromatographic detectors (eg. absorbance, fluorescence). A reagent is continually added to the column effluent so that it reacts selectively with the sample species to form a detectable product, Figure 14. This is the method of choice for detecting heavy metals because they are incompatible with the alkaline regenerant of a suppressed conductivity system. The success of post column derivatisation depends on a constant eluent and reagent flow and an efficient mixing system. This is so that reaction times are fast, background noise kept to a minimum and dead volume low to reduce band broadening effects. Conventional post-column derivatisation techniques utilise a mixing cell, but in 1983 with the advent of membrane technology Dionex introduced the membrane reactor. This is a hollow fibre which is permeable to the reagent and allows for a more homogeneous addition of the reagent than does a mixing cell, Figure 15. This has greatly improved the simplicity and sensitivity of post-column techniques, Figure 16. The post column reagent for heavy metals is (4-(2-pyridylazo)-resorcinol), PAR; this is a metallochromic indicator which will react with most transition and lanthanide metals.

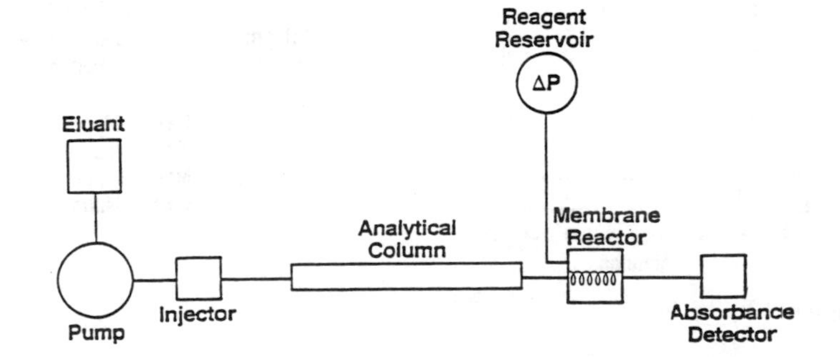

Figure 14

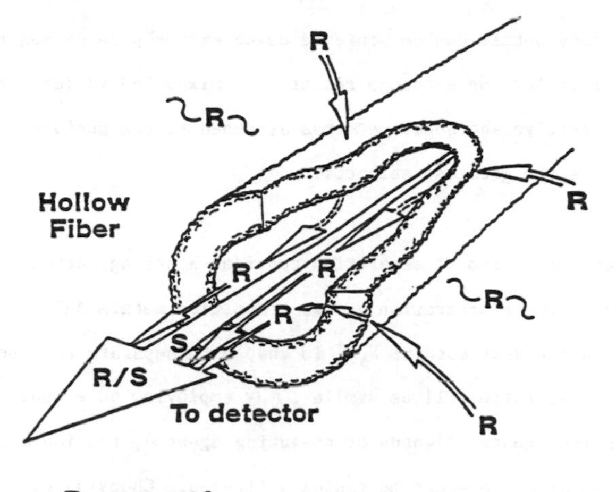

Figure 15 Reagent addition in the membrane reactor

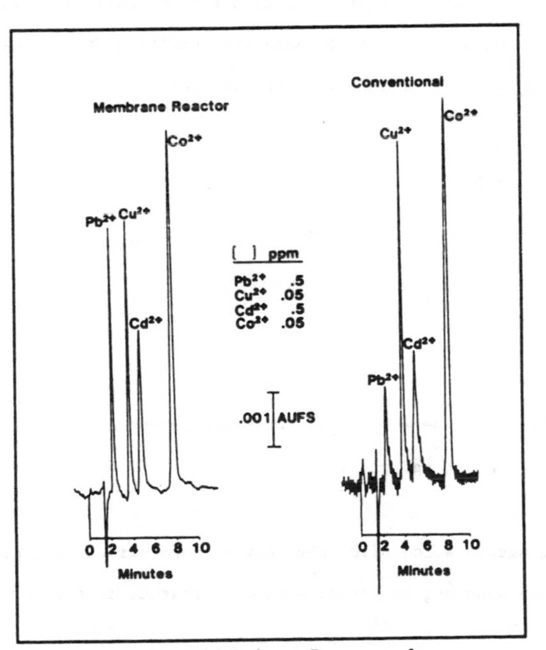

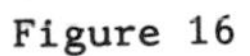

Figure 16 **Signal to Noise Comparison**

Separation of heavy metals can be achieved using either pure cation resins or low capacity pellicular ion exchange resins of a mixed bed variety, that is the resin has both positive and negative sites attached to the surface. The latter has proven to give the greatest selectivity.

To understand the mechanism of separation consider a strong cation exchange resin to be used for the separation of two transition metals An^+ and Bn^+. A monovalent cationic eluent such as H_3O^+ is unable to separate them because their charge to mass ratio will be similar. By employing an eluent that contains complexing agents (ligands or chelating agents), the ion exchange properties of the metal ions can be radicaly altered. Choosing complexing agents for which the formation constants of different metal ion complexes are significantly different allows for rapid and complete separation of most metal ions. The competing equilibria of complex ion formation in the ion exchange process allow for a wide variety of separations of metal ions using anion and cation-exchange resins. This is because the resulting metal complex can be either cationic, neutral or anionic, for example:

$$Mn^+ + L^- \underset{K}{\rightleftharpoons} M(L)_{(n-1)}{}^+$$

$$Mn^+ + nL^- \underset{K}{\rightleftharpoons} M(L)$$

$$Mn^+ + n{+}1L^- \underset{K}{\rightleftharpoons} M(L)$$

The overall charge of the metal will therefore depend on the metal, oxidation state, complexing agent and pH.

The complexing agents which offer the maximum separations on a conventional cation resin are weak organic acids such as tartaric, citric and oxalic.

Optimum selectivity is achieved when two weak acids are present, Figure 17. The retention of the cationic metals occurs when the free metal exchanges onto the chromatographic packing. As the formation of the neutral or anionic complexes increases, retention decreases.

The best overall selectivity is achieved on an anion latex resin using pyridine-2,6-dicarboxylic acid (PDCA) as the chelating ligand. This forms very strong anionic complexes; therefore, the metals are all well separated from each other and the void volume of the chromatographic system, Figure 18. Six metals can be separated in 16 minutes. A modification of the eluent allows 9 metals to be separated in 16 minutes, Figure 19. If oxalic acid is used as an eluent on the same column a mixed retention mechanism occurs and a different

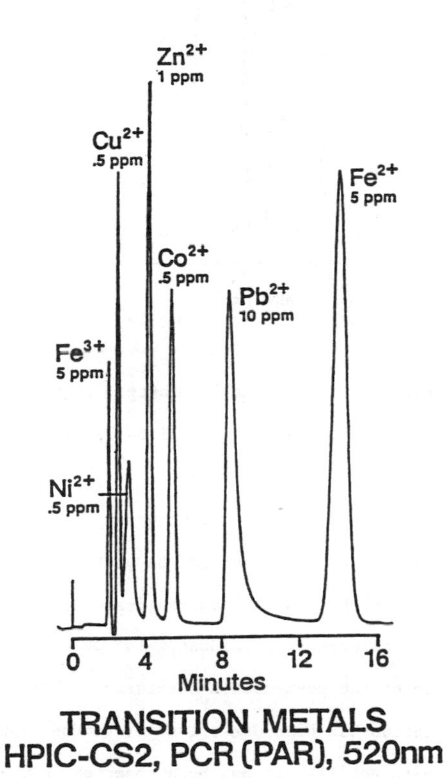

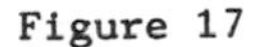
Figure 17 TRANSITION METALS
HPIC-CS2, PCR (PAR), 520nm

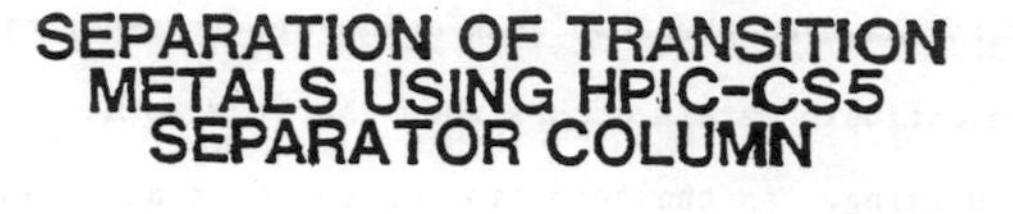

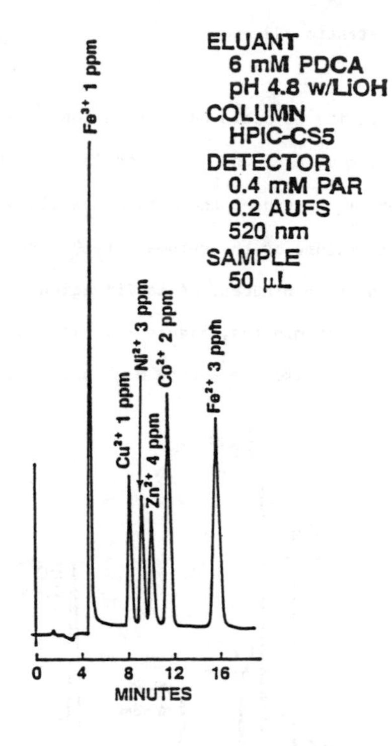

Figure 18

selectivity for 6 metals is achieved in 16 minutes, Figure 20. Sensitivity by all these methods is in the low parts per billion by direct injection.

The scope of metals determined by Ion Chromatography using post column derivatisation with PAR has been limited in the past to the following metals, Mn, Fe, Co, Ni, Cu, Zn, Cd, Pb and the lanthanides. However, it is well known that PAR is a sensitive colourmetric inidcator for many other transition metals. Investigation of the post-column chemistry of PAR with metals has shown that this limitation has been due primarily to the kinetics of complexation of PAR with metals. It has been observed, however, that certain

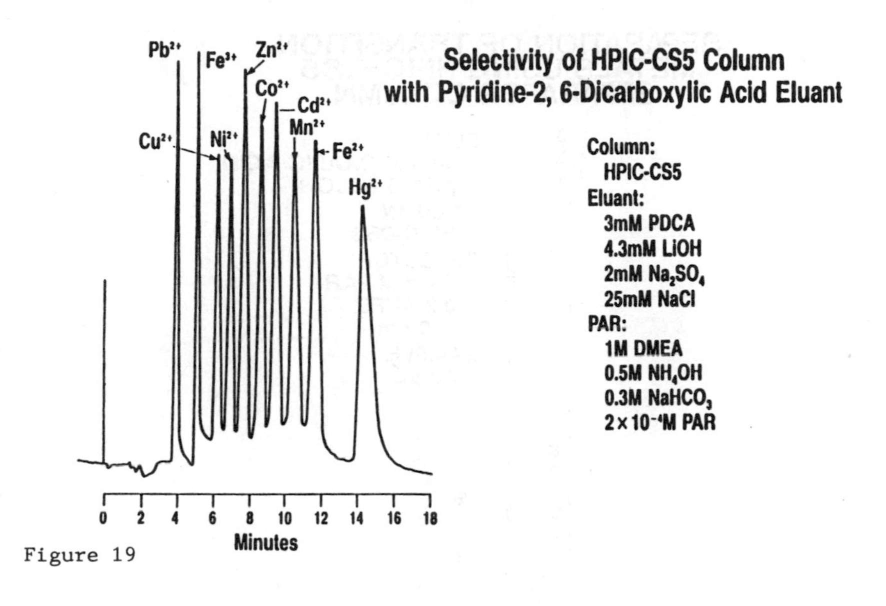

Figure 19

ligands can be added to the post-column solution which enhance the kinetics of complexation of PAR with metals when the eluent chelators are present. It is proposed that the added ligand can solve the metal-eluent chelator complex in such a way as to enhance the kinetics of ligand exchange between the metal and PAR. The result has been that metals, previously masked by the presence of eluent chelators, can be sensitively determined with PAR. Another factor limiting the post-column complexation of PAR with metals has been the pH of the post-column solution. It has been observed that by lowering the pH of the PAR solution from the normal 9.7 will visualise metals previously not detected due to hydrolysis of these metals at a high pH. This is summarised in Figure 21 where Ga, Cr, V and Hg are now detectable.

One of the important aspects of this type of separation is speciation. It has already been shown that Iron II and III can be separated along with vanadium IV and V using the PDCA eluent. Figure 22 shows the speciation of Tin II and IV along with lead. This was developed in order to achieve the rapid separation

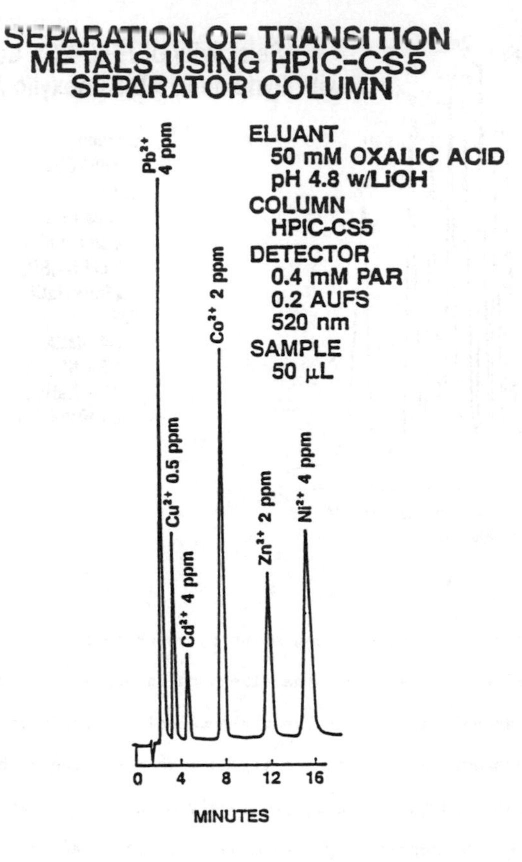
SEPARATION OF TRANSITION
METALS USING HPIC-CS5
SEPARATOR COLUMN
ELUANT
50 mM OXALIC ACID
pH 4.8 w/LiOH
COLUMN
HPIC-CS5
DETECTOR
0.4 mM PAR
0.2 AUFS
520 nm
SAMPLE
50 μL
Pb2+ 4 ppm
Cu2+ 0.5 ppm
Cd2+ 4 ppm
Co2+ 2 ppm
Zn2+ 2 ppm
Ni2+ 4 ppm
0
4
8
12
16
MINUTES

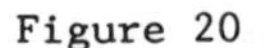
Figure 20

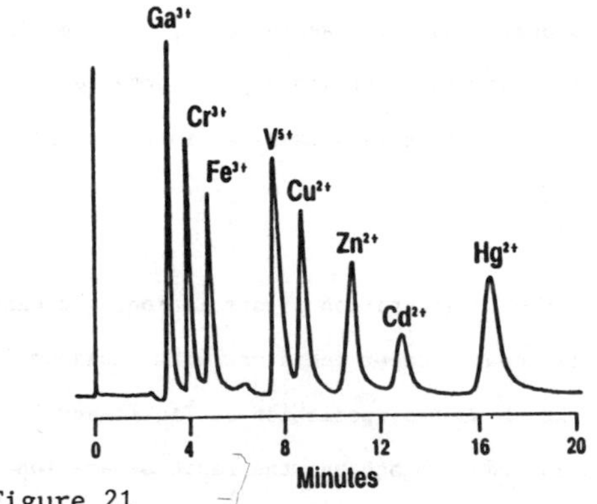
Determination of Mercury(II)
Column:
HPIC-CS5
Eluant:
6mM PDCA
8.6mM LiOH
PAR:
0.5M Na2HPO4
2×10-4M PAR
Detector:
VIS
520nm
Hg2+ — 20ppm
Ga3+
Cr3+
Fe3+
V5+
Cu2+
Zn2+
Cd2+
Hg2+
0
4
8
12
16
20
Minutes

Figure 21

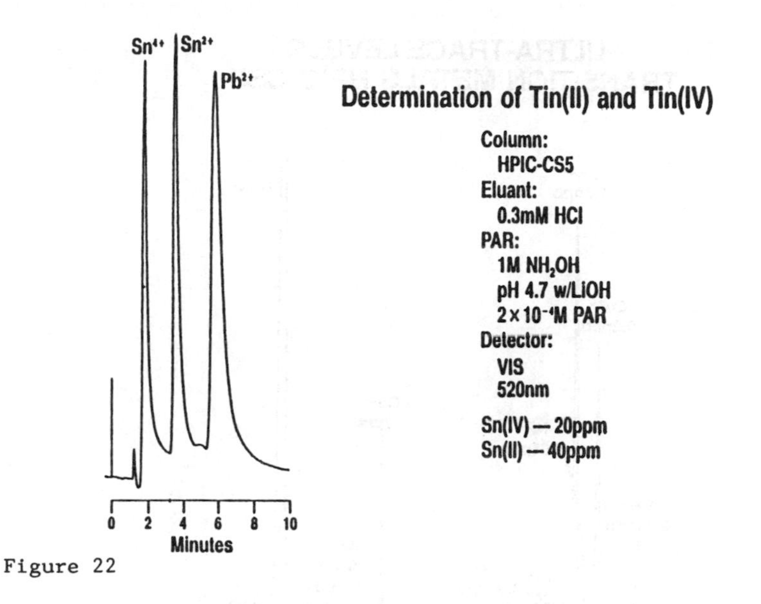

Figure 22

of the major constituents of a tin-lead solder bath. The separation was achieved using an HCl eluent. Detection limits are approximately 0.05ppm by direct injection.

Figure 23 represents a very powerful technique within all forms of Ion Chromatography, that of trace enrichment. Samples containing low levels of ions can be concentrated up by passing a few mls through an appropriate short precolumn which is later injected into the system and the concentrated ions removed by the buffer and analysed in the normal way. Note that parts per trillion concentrations are being monitored using this method. Detection limits are two to three orders of magnitude better than using conventional spectroscopic techniques.

Another technique similar to trace enrichment is matrix elimination which is used for highly corrosive or concentrated matrices such as acids and salts. Figure 24 shows low levels of metals in concentrated sulphuric acid. The

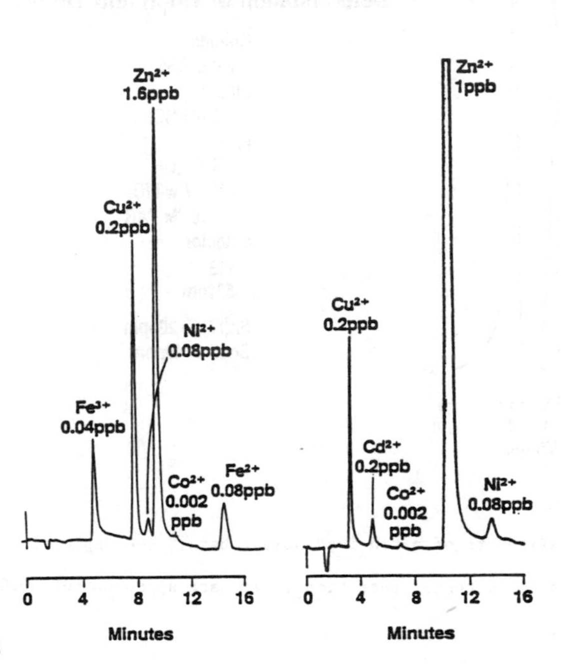

Figure 23

sample is first injected onto a precolumn which is then rinsed with metal free deionised water to remove the excess anions. Using conventional spectroscopic techniques these levels in concentrated acids cannot be determined without extensive pretreatment.

Having the ability to change selectivity becomes a powerful tool when difficult samples are encountered as can be seen from the analysis of an electroless nickel plating bath using the oxalate eluent. This permits the analysis of the trace ions of interest; Pb, Cd, Cu in the presence of 1000ppm of nickel, Figure 25.

Lanthanide metals also react with PAR and can be separated using either anion or cation exchange. Step gradients have been used in the past, but with the

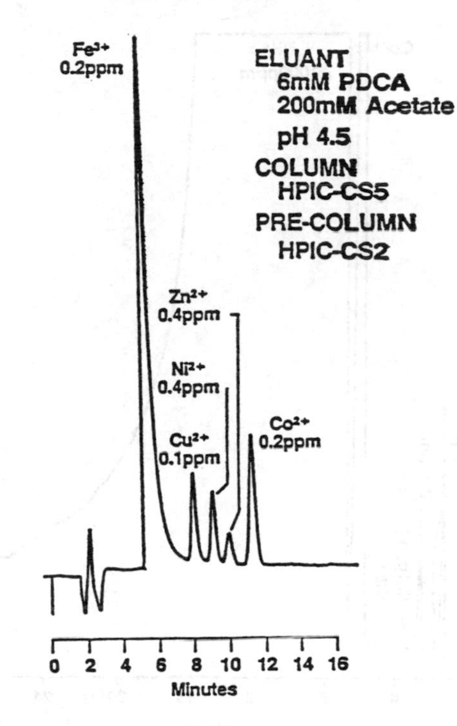

Figure 24

advent of gradient capability to the range of Dionex equipment it is possible to separate 14 lanthanides by a linear gradient in under 20 minutes, Figure 26.

Other post column reagents have been used for specific metals; for example, diphenyl carbazide can be used for the very sensitive detection of chromate, an important environmental polutant. Less than 1ppb can be detected by direct injection (50μl). A simple pre-derivatisation of the same sample with PDCA buffer allows detection of CrIII and thus speciation of CrIII and CrVI, Figure 27.

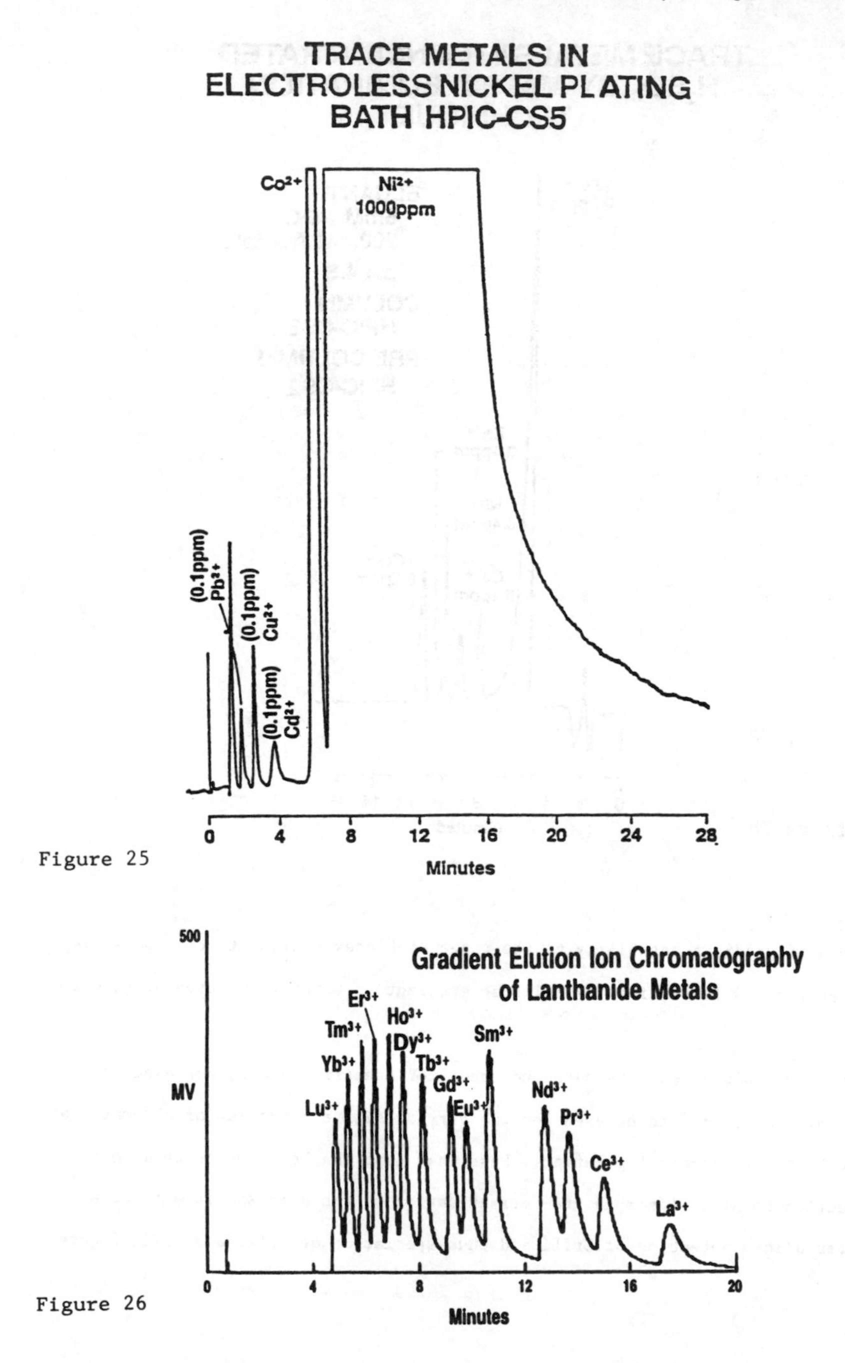

TRACE METALS IN
ELECTROLESS NICKEL PLATING
BATH HPIC-CS5
Co²⁺
Ni²⁺
1000ppm
(0.1ppm)
Pb²⁺
(0.1ppm)
Cu²⁺
(0.1ppm)
Cd²⁺
0
4
8
12
16
20
24
28
Minutes
500
Gradient Elution Ion Chromatography
of Lanthanide Metals
MV
Er³⁺
Tm³⁺
Ho³⁺
Dy³⁺
Sm³⁺
Yb³⁺
Tb³⁺
Gd³⁺
Nd³⁺
Lu³⁺
Eu³⁺
Pr³⁺
Ce³⁺
La³⁺
0
4
8
12
16
20
Minutes

Figure 25

Figure 26

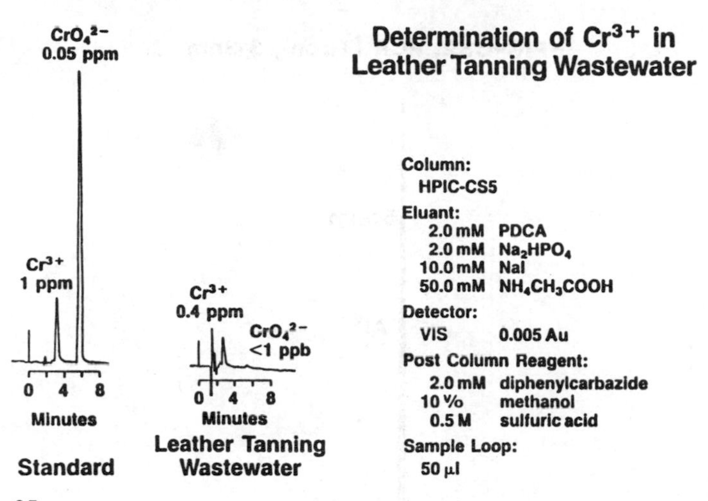

Figure 27

Tiron indicator is a very specific and sensitivie post column reagent for the analysis of Aluminium at 313nm, Figure 28.

When performing analysis of transition metals it is important that the chromatography system is metal free. Dionex Ion Chromatography systems are totally metal free in the eluent flow path. Figure 29 illustrates this requirement.

In conclusion, post column derivatisation offers the following advantages over other analytical methods:

- Sensitivity, ppb levels by direct injection and even lower by trace enrichment techniques.

- Multi-element technique with variable selectivity.

- Speciation of important oxidation states.

- Matrix elimination technique for complex samples.

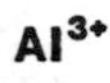

Figure 28 Determination of aluminium by cation exchange

SUMMARY

Ion Chromatography has since its beginning found increased use as an alternative or compliment to conventional methods for the determination of metals. Through the use of ion exchange separations in a variety of ways and in combination with either suppressed conductivity detection or post column reation/visible detection, it is possible to detect a wide range of metals in a variety of oxidation states.

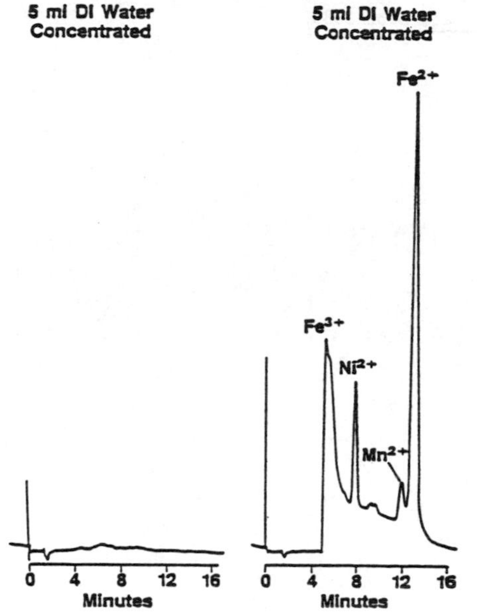

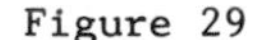
Figure 29 Dionex Metal-Free QIC Pump Conventional Stainless Steel HPLC Pump

The speed and versatility of the system has made it todays fastest growing analytical technique and the method of choice for the analysis of inorganic anions and cations.

REFERENCES

1. SMALL, H., STEVENS T.S., AND BAUMAN W.C., Anal.Chem. 47,1801 (1975)

2. STEVENS, T.S.,JEWETT, G.L., BREDEWEG, R.A., Anal.Chem. 1982, 54,1206

3. T.S. STEVENS, J.C. DAVIS and H. SMALL, Anal.Chem. 54,1488 (1981)

4. J. STILLIAN, LC Mag, Sept 1985

Recent Advances in Graphite Furnace Analysis

By D. Agness

PERKIN-ELMER LIMITED, POST OFFICE LANE, BEACONSFIELD, BUCKINGHAMSHIRE HP9 IQA, UK

Introduction

Since the introduction of Graphite Furnace Analysis the technique has shown itself to be very sensitive for a wide range of elements in various matrices. Over the years however, many interferences have become apparent and in many instances no way of overcoming the problems could be found, apart from standard additions.

As the method of standard additions is very time consuming and, in the past, very operator dependent a considerable amount of work has been undertaken to try and identify the interferences and find ways to over come them. Many of these advances have taken place over the past few years and are still not fully accepted in some laboratories, possibly due to a lack of understanding on the part of the user.

The aim of this paper is to try and explain some of the changes that have taken place and their relevance to low level metal analysis. Many of these changes on their own however will not overcome the interferences completely. It is therefore important for the analyst to understand that to totally eliminate interferences one is required to use all these developments together and not in isolation as many people do.

Advances that have taken place are :-

1) Area measurement of analyte signal
2) Base line off-set correction
3) Fast heating rates (for atomisation of analyte)
4) L'VOV Platform
5) Tube Introduction
6) Matrix modifiers and alternate gases
7) Minimal temperature step from sample preparation (char) to atomisation step

8) Background Correction
9) Graphics for method development
10) Furnace programming
11) Autosampler programming
12) Solid Sampling

Signal Measurement

In the early days of furnace analysis it was traditional to measure the analyte absorbance using peak height. The main reason for this is due to the fact that in the past signals were output to a recorder, making peak height measurement the choice. As electronics improved and digital display became a standard feature on all instruments peak height measurement along with peak area became available. However on many instruments area was still not viable due to the low speed of the electronics compared with the speed of the analytical peak.

However with the introduction of fast graphical displays (VDU's) showing the signals in their untreated form, true peak shape could be studied and accurate area measurement became possible. Normal pen recorders are mechanical devices and dampen the signal as a matter of course, since full scale deflection of the pen requires half a second.

Figure 1 shows an example of lead injected onto a pyro coated tube with L'VOV platform in place. The two peaks are of the same sample but the second broader peak occurs after the tube has had 250 firings. Measuring the peak height over this period the signal dropped by 28% but measuring peak area a variation of only 2% was recorded over the same number of firings. It is inevitable that any tube, whatever type, will age with use but by using peak area measurement the change in signal intensity can be reduced, resulting in the ability to analyse samples in long runs with minimal re-standardisation.

Figure 2 shows an example of cadmium, the first peak being the standard and the second peak cadmium in a potable water sample, both containing the same amount of cadmium. If peak height was used to measure the concentration approximately 20% suppresion would result.

FIGURE 1

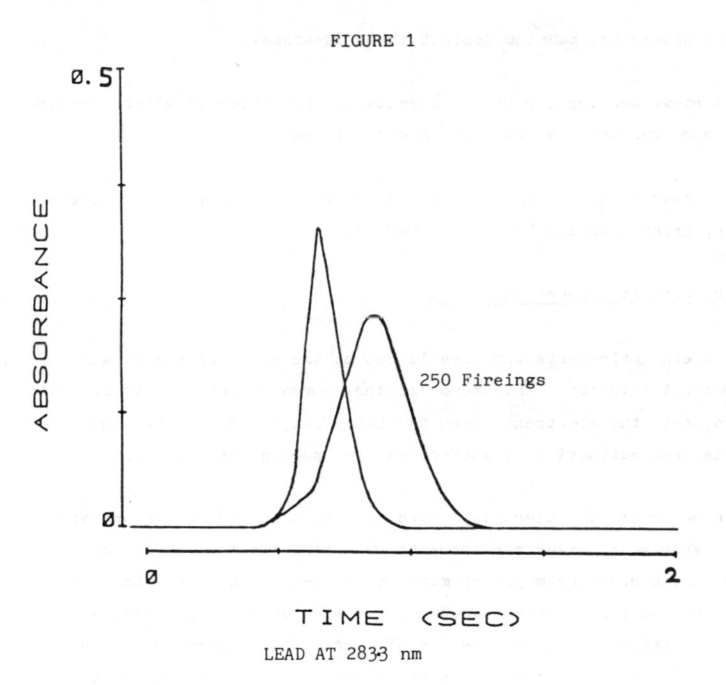

LEAD AT 283.3 nm

FIGURE 2

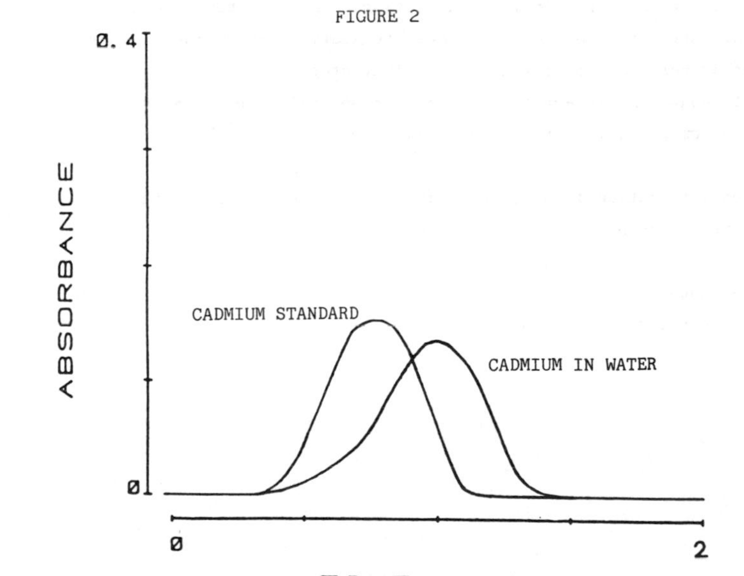

Peak area showed the cadmium content to be identical.

Figure 3 shows working curves for cadmium determination in water, showing that area measurement is important along with other conditions.

Figure 4 shows a similar example with lead in blood, from a study carried out by E. Pruskowska and W. Slavin (Ref 1).

Base Line Off-Set Correction (B.O.C.)

On all modern instruments the zero is set at the start of a analysis using a blank solution. This zero setting, normally referred to as Autozero, sets the electronic zero and blank sample zero. The blank signal is then automatically subtracted from subsequent readings.

BOC is used to set the electronic zero just before reading the analyte signal. When peak height measurements are used, small uncorrected changes in the baseline maybe of minor consequence. However, the effect of baseline changes is more pronounced for even small displacements when area integration is used as the displacement is summed over the complete integration period. This applies to both double beam system and more importantly to single beam systems. The BOC should be set after the matrix is removed and before atomisation takes place. This is normally performed by extending the char step. The BOC time should be set to equal the integration period. On modern instruments this is done automatically just before atomisation.

When area integration is used Base Line Correction must be performed before every reading as a matter of course.

Figure 5 shows a typical furnace program using BOC on a non VDU system where it is not performed automatically.

Figure 3

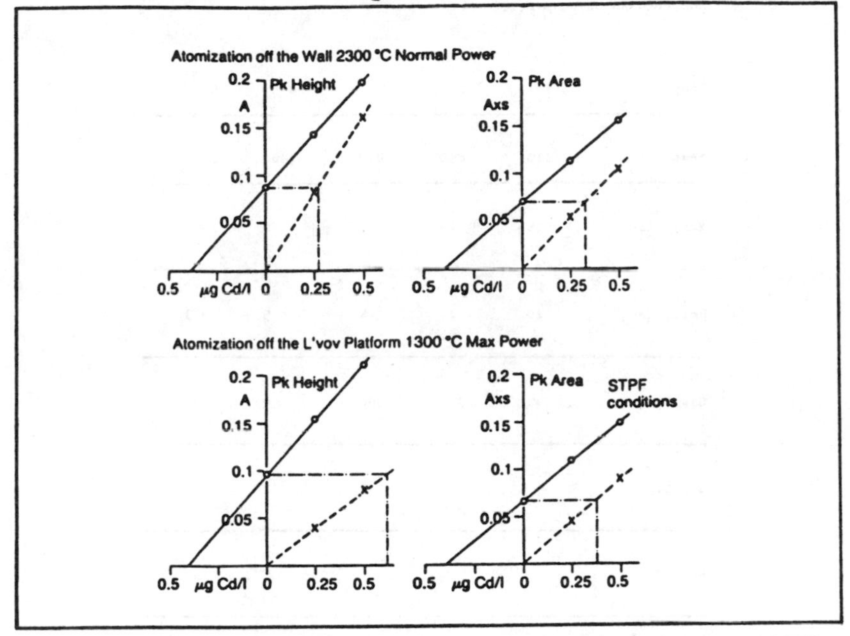

Different working curves for the determination of Cd in waste water. Only under real STPF conditions the sample can be measured directly against aqueous standards.

FIGURE 4

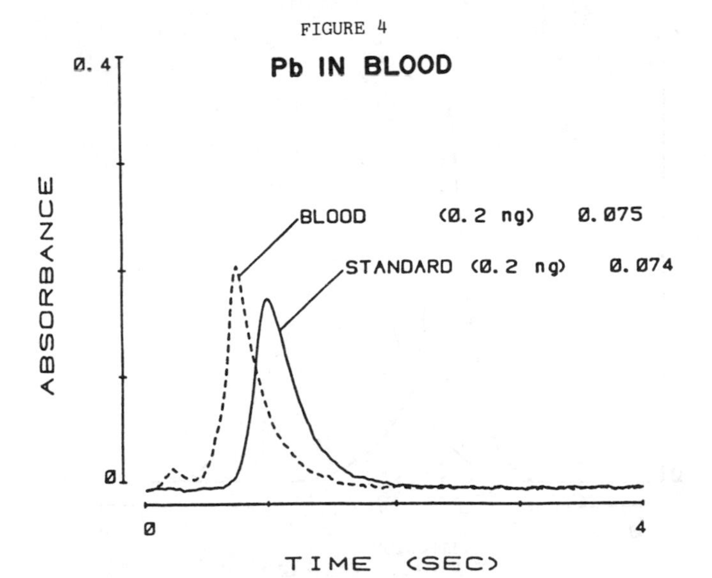

FIGURE 5

HGA PROGRAM USING B.O.C.

Step	1	2	3	4	5
Temp °C	110	850	850	1800	2650
Ramp, Sec	15	10	1	0	1
Hold, Sec	20	20	5	5	3
Gas Flow	300	300	300	0	300
B.O.C.			•		
Read				•	

Step 3 - Base Line Correction Stage

FIGURE 6

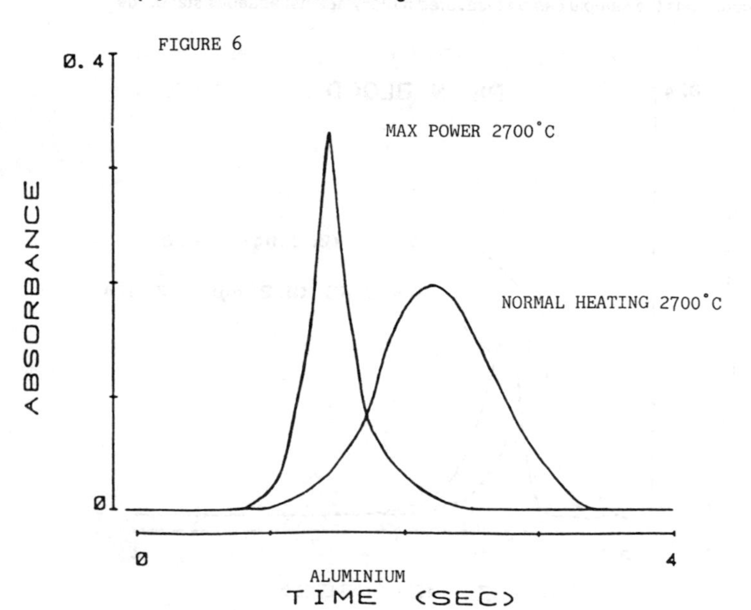

Fast Heating Rates (Maximum Power)

With maximum power heating the graphite tube is heated to the preselected temperature at a speed faster than 2000°C/sec.

Slower heating rates means that high tube temperatures have to be used to atomise the element of interest. A shorter tube life and for some elements reduced sensitivities is the result.

Figure 6 shows the atomisation of Al using slow heating and maximum power heating. It can be seen that slower heating rate gives lower sensitivities and requires this temperature to be maintained longer to atomise all the Al. Even though in this example the same atomisation temperature is required long tube life is still possible: see Figure 7 Ref 2.

However with elements like lead shown in Figure 8 it is possible to reduce the atomisation temperature significantly without any loss in sensitivity by using maximum power heating.

Maximum power heating also allows the tube temperature to reach thermal equilibrium before the element starts to atomize, and thus prevent double peaking and the sample atoms from being carried out of the light beam by expanding gas.

LVOV Platform requires max power heating to work correctly.

LVOV Platforms

LVOV Platforms are possibly one, if not the most important development in graphite furnace technology in the past years. Platforms delay the atomisation until the temperature of the inert gas in the graphite tube is in equilibium with the wall temperature.

FIGURE 7

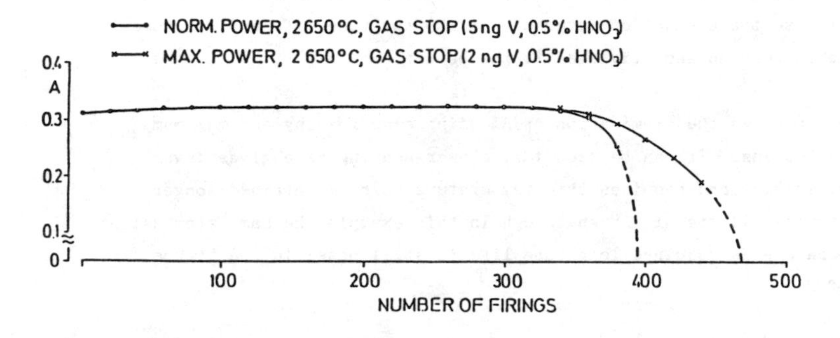

FIGURE 8

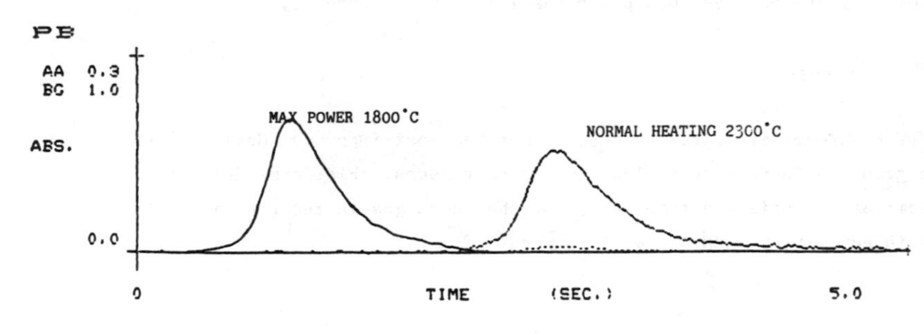

LEAD AT 283.3 nm

Figure 9 shows a typical platform and its position in the graphite tube. The early design of platform were found to have a few restrictions with respect to sample volume and the pipetting of some viscous samples such as urine and blood. The platform was therefore redesigned to make it possible to deposit larger samples of a more viscous nature but without changing the outer physical dimensions of the platform.

Figure 10 illustrates these changes. The sample area is now wider (3mm compared to 2mm) and the inner wall is no longer rectangular to the surface. The bottom surface itself is no longer flat but slightly V shaped. These changes gave an increase of platform volume of approx. 100%. The maximum useful volume is now 50µl for aqueous samples. For samples like milk and urine up to 40µl of sample can now be pipetted onto the platform. If this volume is still too small to provide a significant signal repeat injections can be made after drying and/or sample pre-treatment to increas sensitivity.

The analytical lifetime of platform have been tested using 5 to 20% nitric using analytical conditions for copper. Approximately 1000 firings were made before the tube broke, but the platform still appeared to be more or less unaffected (see Figure 11).

6) Matrix Modifiers and Alternate Gases

Matrix modifiers were first suggested by Ediger in 1975 (3); since this time many people have expanded the idea and many useful modifiers have been found.

Modifiers are a very important aspect for interference free trace metal analysis, and have made many determinations possible by direct injection. The materials used as the modifier may change the chemical nature and the physical properties of the element under study, or in many cases the matrix.

The modifier can either be added to the sample directly or more conveniently added to the sample in the graphite furnace. This can be performed automatically by autosamplers now available, thus making the analysis simpler and reducing the possibility of contamination.

FIGURE 9

L'VOV PLATFORM IN GRAPHITE TUBE

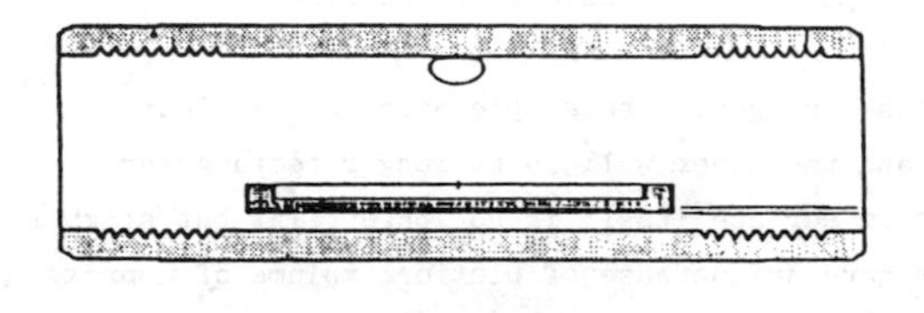

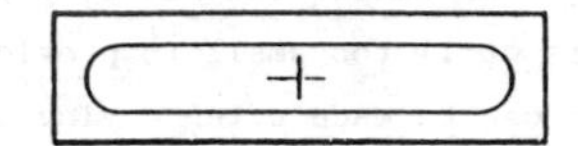

L'vov Platform, manufactured from solid pyrolytic graphite, mounted inside of a graphite tube.

FIGURE 10

CROSS SECTION
OF AN EARLIER L'VOV PLATFORM

CROSS SECTION
OF A RECENT L'VOV PLATFORM

FIGURE 11

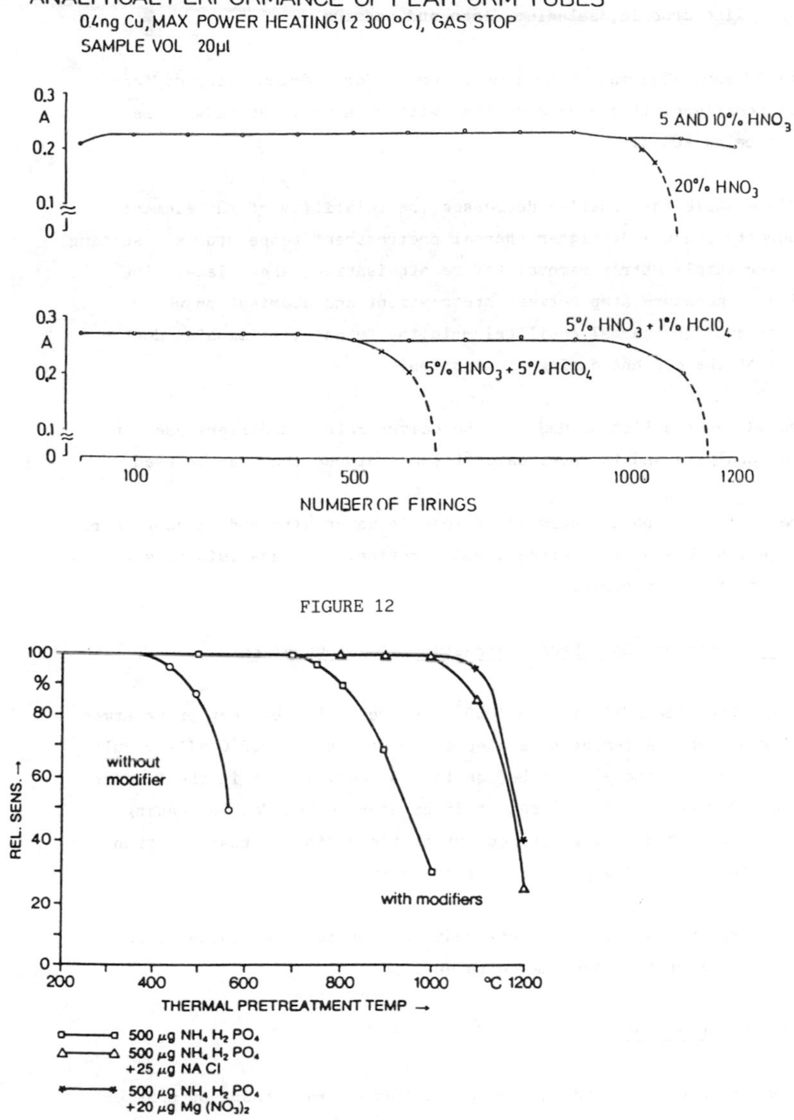

Thermal pretreatment curves for cadmium using different matrix modifiers

In many furnace determinations it is difficult to remove the sample matrix without loss of the element of interest; this is typical with elements like arsenic, selenium, lean and cadmium.

Figure 12 shows Thermal pretreatment curves for cadmium using different matrix modifiers, it can be seen that without a modifier Cadmium is lost at below 400°C.

In this example the modifier decreases the volatility of the element and permits the use of higher thermal pretreatment temperatures resulting in better sample matrix removal before atomisation takes place. The smaller temperature step between pretreatment and atomisation which is important for thermal equilibrium in the furnace, eliminates double peaking of the element during atomisation.

Figure 13 shows a list of many of the common matrix modifiers used in routine analysis and the pretreatment temperatures that can be used.

Figure 14 is a graphical example of lead in water with and without matrix modification showing why, without modification, this analysis is very difficult if not impossible.

Minimal Temperature Step From Pretreatment to Atomisation

A temperature step of less than 1000°C is preferable between pretreatment and atomisation. A temperature step of greater than 1000°C will result in the release of the element before thermal equilibrium in the furnace has been reached. This will result in broader peaks, double peaking and the sample atoms being carried out of the radiation beam by expanding gases, resulting in a suppression of the signal.

Using max power heating and matrix modification together temperature steps can be kept to a minimum quite easily.

Background Correction

Many samples when analysed in a graphite furnace may absorb or scatter the radiation from the source. This is due to salt particles, smoke etc which results in an increase in the total signal and the reporting of high results.

FIGURE 13

Matrix Modifiers and Related Thermal Pretreatment Temperatures

Element	Modifier	Pretreatment Temp. °C	Element	Modifier	Pretreatment Temp. °C
Al	$Mg(NO_3)_2$	1700	Ni	$Mg(NO_3)_2$	1400
As	Ni	1300	P	$La(NO_3)_3$	1350
Be	$Mg(NO_3)_2$	1500	Pb	$NH_4H_2PO_4$	950
Cd	$NH_4H_2PO_4$	750	Sb	Ni	1000
Co	$Mg(NO_3)_2$	1400	Se	$Cu/Mg(NO_3)_2$	1100
Cr	$Mg(NO_3)_2$	1650	Sn	$HNO_3 + NH_4OH$	1000
Fe	$Mg(NO_3)_2$	1450	Te	Cu or $(NH_4)_2Cr_2O_7$	900
Hg	$K_2Cr_2O_7$ or Te	250	Tl	H_2SO_4	750
Mn	$Mg(NO_3)_2$	1400	Zn	H_3PO_4 or $NH_4H_2PO_4$	900

FIGURE 14

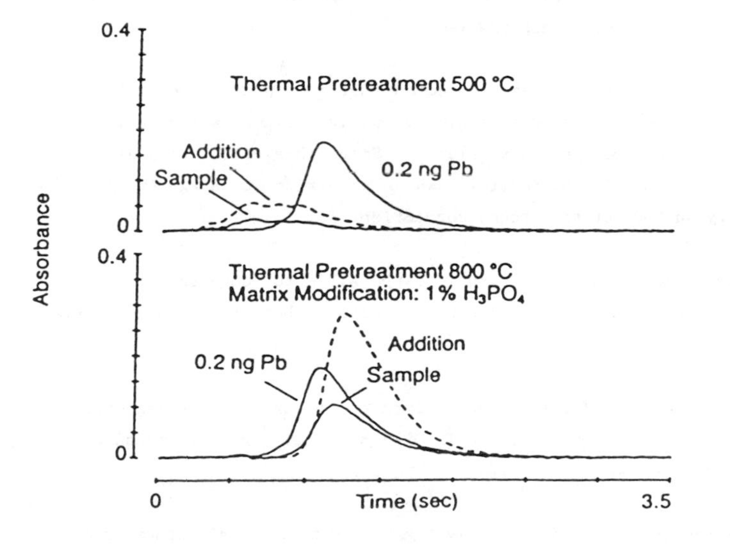

For all furnace work simultaneous background correction should be used. By far the most commonly used form of background correction is performed using a continuum source corrector (Deuterium arc).

A Deuterium arc background corrector, although very effective, has many limitations. It only works well in the UV range 190 - 340 nm. A tungsten source is required for the visible region 300nm to 850nm, for example which is only available. Background is in a few instruments generally only totally corrected if the background signal is less than 1.00 absorbance unit, although at some wavelengths higher background signals can be corrected. Only scattering and molecular absorption that is within the spectral band can be corrected for accurately. Any fine structured absorption can result in errors due to over or under correction.

If a rapidly changing background is present as shown in figure 15 total correction is not performed even at low background levels.

This is due to the way in which the background signal is measured, normally only on one side of the analyte peak. Some newer background correction systems however do sample the background on either side of the analyte, overcoming this problem.

In the late 19th century Zeeman discoved that if a sodium spectral emission line was subjected to a high magnetic field it was polarized into parallel and perpendicular planes. Since this effect only effects the element atoms and not particles which give rise to background it is ideally suited for background correction.

With Zeeman background correction it is possible to compensate for background absorption up to 2.00 absorption units in both the UV and Visible region.

Figure 16 shows the effect on the analyte component with the magnetic field on and off. When the field is OFF the total signal is measured, atomic and non-atomic.

When the magnetic field is on, only the non-atomic signal is measured, as the sigma components of the atomic line are shifted away from the resonance line. The absorption component the resonance line is not measured due to a polarizer in the optical system of the spectrophoto-

FIGURE 15

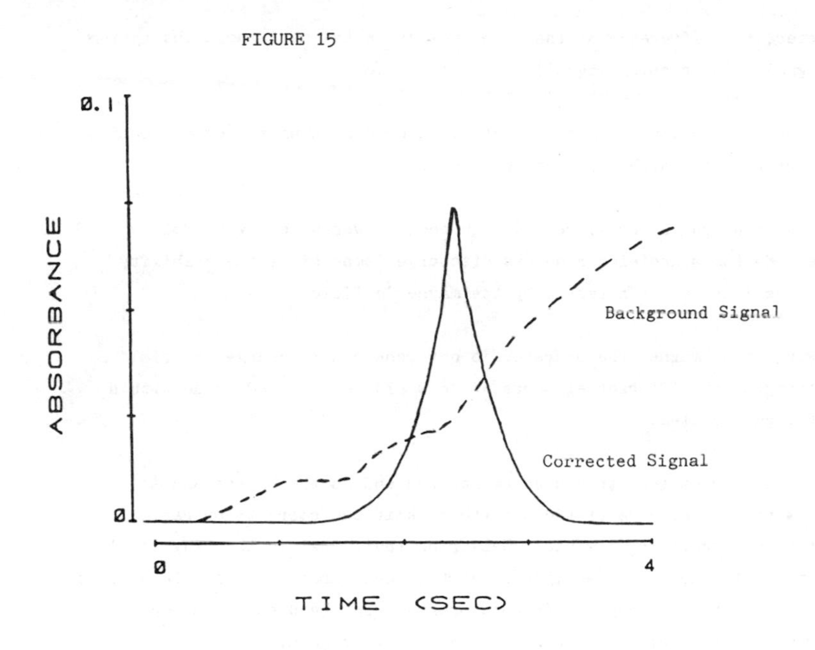

FIGURE 16

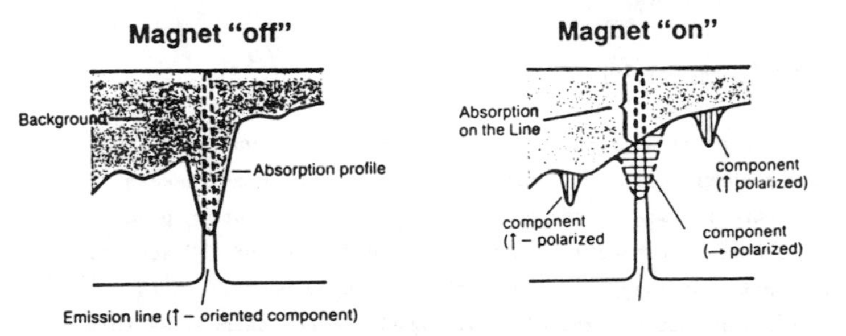

meter; the difference of the two readings is the net atomic absorption signal (Correction signal).

Zeeman correction has many advantages over Continuum source correction apart from the higher correction levels.

No special radiation source is required; it works well with Hollow Cathode Lamps and electrodeless discharge lamps since the stability of the lamp is not affected by the magnetic field.

Using an AC magnet the operator is not required to change the field strength for different elements as he would be required to do with a DC magnet system.

It can correct for structured background and as the background is measured either side of the analyte peak it can correct for rapidly changing background signals. Measuring the signal in this way it also compensates more accurately even at the lower absorbance levels of background. Shorter graphite furnace programs can be used as the correction is more efficient, thus increasing sample throughput. There is a slight loss in sensitivity for some elements but this in no way affects the final detection limit possible with graphite furnace analysis.

It can be seen in the literature that Zeeman correction shows considerable practical advantages over the Continuum source type of background correction and more and more people are becoming familiar with the Zeeman system.

Graphics for Method Development

Method development of a graphite furnace program was always difficult using a strip chart recorder. This was due to the slow response of all chart recorders, making the recording of fast atomisation peaks difficult to monitor correctly. With the introduction of VDU screens and fast graphics software this problem has been overcome. It is important however that the graphics displayed are not damped, as this gives rise to a similar problem as a strip chart recorder. It is now possible not only to monitor the true analyte peak but also the background peaks, displayed in real time on a single screen.

Figures 17 and 18 show an example for lead. The only change in the program was an increase in the pretreatment from 450°C to 950°C and the use of matrix modification. Note the change in the peak retention time and the almost complete removal of the background signal. Without a VDU system these changes would not have been apparent. Many of the examples shown in this paper would have been impossible to illustrate without a fast graphics system.

By seeing the peaks in their true form many interference problems can be explained and compensated for quite easily.

Furnace Programming

In the early days of Graphite Furnace systems it was only possible to change the temperature three or four times within a program. This has shown itself to be a restriction on obtaining a good reproducible result for many applications. Modern furnaces permit the temperature, ramp and hold time to be changed up to nine times within a program. It has become apparent that to use a two step dry will improve the drying nature of may samples, like blood, urine, seawater and thus improve reproducibility. This controlled increase in temperature permits the release of the combustible products slowly and in order of volatility. A tube cleaning step is also useful in most programs, making sure the tube is clean between each sample, resulting in longer tube life and no sample memory problems.

Today typical graphite furnace programs can have six or more steps, this however does not result in longer analysis time just better results. An example of a six step program can be seen in Figure 19 for the analysis of arsenic in horse urine.

It is even possible to program the Graphite Furnace through the VDU screen and to store the complete method along with instrument and auto-sampler parameters for rapid recall when required.

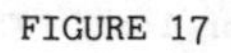

FIGURE 17

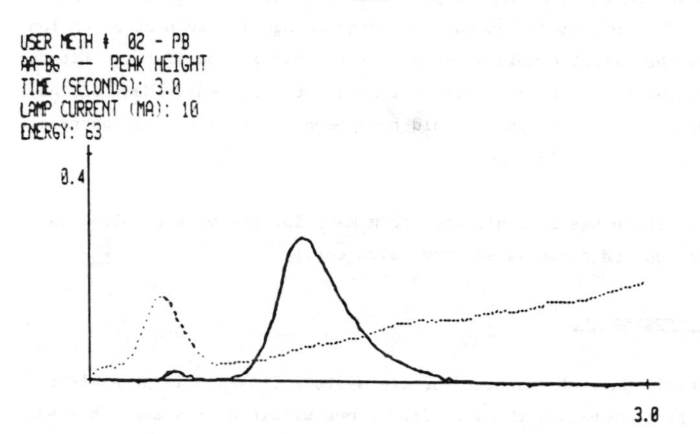

FIGURE 18

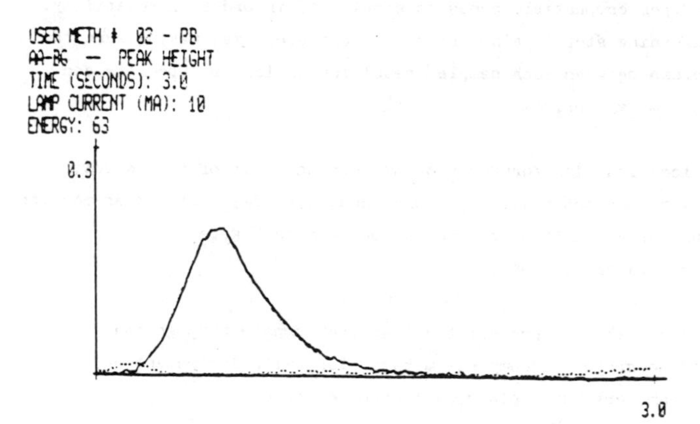

FIGURE 19

Step No	Furnace Temp	Ramp	Hold	Internal Flow	Read
1	80	2	15	300	
2	120	10	25	300	
3	650	10	20	300	
4	900	10	10	300	
5	2150	0	3	0	*
6	2650	1	3	300	

MATRIX MODIFICATION USED.

FIGURE 20

PROGRAMMING MODE AUTOSAMPLER STD. METH #

SOLUTIONS	LOCATION	VOLUME	BLANK VOLUME
BLANK	00		20
STANDARD 1	01	05	15
STANDARD 2	01	10	10
STANDARD 3	01	15	05
STANDARD 4	01	20	--
STANDARD 5	02	05	15
STANDARD 6	02	10	10
STANDARD 7	02	15	05
STANDARD 8	02	20	--
RESLOPE	--	20	--
MATRIX MODIFIER 1	40	10	--
MATRIX MODIFIER 2	39	10	--
SAMPLE 03 TO 20 WITH MODIFIER 1 + -		20	--
SAMPLE 21 TO 38 WITH MODIFIER - + 2		20	--

RECAL LOCATIONS: 10 : 20 : 30 : -- : -- NUMBER OF INJECTIONS: 01

Autosampler Programming

Providing the furnace program is correct the introduction of the sample is the final area where reproducibility is influenced. In the past the reproducibility of any furnace method was limited by the manual injection system (the operator). Not only did he or she have to dispense the volume correctly but also the position of the sample in the tube had to remain constant.

Volume, sample position, speed of delivery all influence the precision of the results.

With the introduction of the first furnace autosampler in 1976 the burden of manual pipetting was removed. Autosamplers can accurately pipette volume from 5µl to 100µl and also peform multi-injection sequences, making it possible to lower detection limits even further. Modern autosamplers can inject the Blank, Standards, Modifiers and Samples in any order and volume the operator requires. This makes it possible to do multi-standard calibrations with only one standard, just by changing the volume injected, final volume always being constant with the addition of bland. The method of additions can also be automated for up to eight additions. Modifiers concentrations can be varied to find the correct amount, overnight runs can be performed to increase sample throughput and laboratory efficiency.

Figure 20 shows a VDU screen display used to program an autosampler in any way the operator requires. The example shown is for eight standard calibration.

Solid Sampling

Solid sampling by graphite furnace has always been practical, but until the introduction of Zeeman background correction one draw back was the larger background levels some samples exhibit. This of course is no longer a problem.

With the latest design of solid sampling tubes (Figure 21) the centre of the tube can be removed. This allows the operator to weigh the sample directly onto a small pyrolytic graphite cup, then re-insert it into the graphite tube. The cup acts like a L'VOV Platform, reducing or eliminating many interferences.

Matrix modifiers and/or acid can then be added directly into the tube by the autosampler. In this way many solids can be analysed directly against aqueous standards, thus overcoming the need for solid reference materials. With the solid sampler the best results are generally acheived when the sample is pulverized or homogenized and weighed on a five place balance. For the introduction of powders a sampling tool is available to make the task easier.

Figure 22 shows the direct analysis of cadmium in tobacco; the results agreed with the analysis of the digested sample.

Tube Introduction

The method of mounting of the graphite tubes in any furnace system is a very important feature. The contact between the tube and cones (contact cylinders) must remain constant for all tubes, otherwise furnace temperatures will not be reproducible from tube to tube.

On some furnace systems the operator must unscrew the contact cylinders to remove the old tube and then reclamp the new tube in place. This can lead to a bad contact between the two graphite surfaces. Should the pressure on the tube be too great the tube may even break on heating.

As the operator will be required to change tubes on a regular basis the system employed to hold the tube should be efficient but easy to use.

The most efficient way would be for the furnace to control the contact pressure. On many new systems this is done by pneumatic control of the furnace head. The operator simply opens the furnace by releasing the gas pressure, places in a new tube of any type and reinstates the gas supply to the heads. The whole operation takes less than 20 seconds to complete.

FIGURE 21

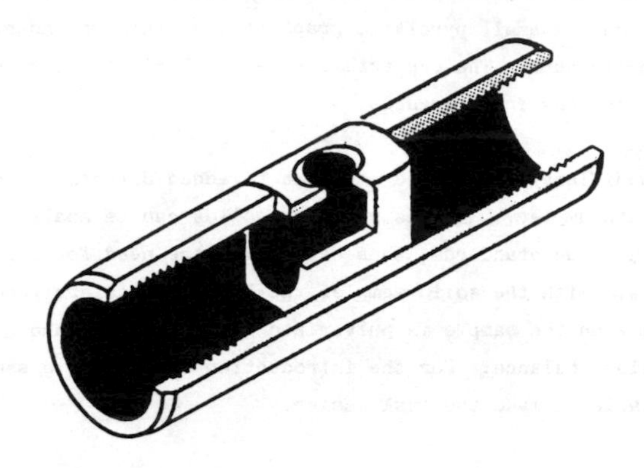

HGA Solid Sampling Accessory

FIGURE 22

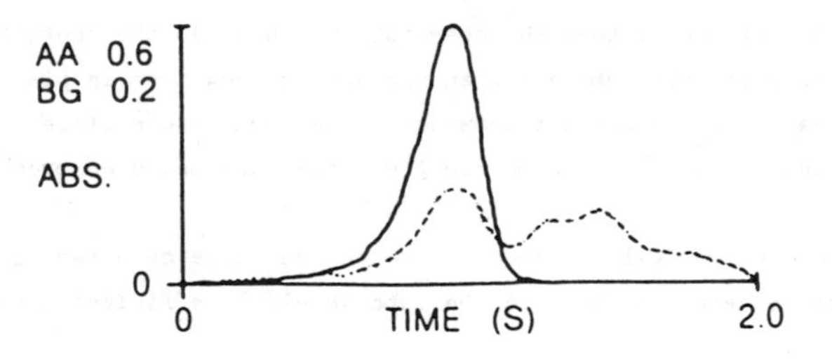

Determination of (1 27 μg/g) Cd in tobacco using direct solid sampling (results are in agreement with analysis of decomposed sample)

The autosampler if used is also still in alignment with the tube, resulting in minimum delay in analysis time.

Conclusion

With the improvements now available graphite furnace has become an excepted technique in many laboratories, for routine analysis of a wide range of samples were low level metal content is required.

Once the methods have been developed it is possible with modern computer controlled systems for the analysis to be carried out by relatively untrained staff, as all the conditions can be recalled from memory.

References

1) Biochemical Analyses with S.T.P.F. and Zeeman Background Correction
E. Pruszkowskz and W. Slavin
A. S. Application Study 684

2) Improvements in Graphite Furnace Tubes
U. Voellkopf and Z. Grobenski
A. S. Application Study 688

3) R. D. Ediger, At absorpt Newslett.14, 127 (1975)

4) Interference - free Trace Metal Determination. Perkin-Elmer

5) Perkin-Elmer LA6 Notes